T0264946

Nonparametric Statistical Tests

A Computational Approach

Markus Neuhäuser

CRC Press
Taylor & Francis Group
Boca Raton London New York

CRC Press is an imprint of the
Taylor & Francis Group, an **informa** business

A CHAPMAN & HALL BOOK

CRC Press
Taylor & Francis Group
6000 Broken Sound Parkway NW, Suite 300
Boca Raton, FL 33487-2742

First issued in paperback 2017

© 2012 by Taylor & Francis Group, LLC
CRC Press is an imprint of Taylor & Francis Group, an Informa business

No claim to original U.S. Government works

ISBN-13: 978-1-4398-6703-7 (hbk)
ISBN-13: 978-1-138-11410-4 (pbk)

Visit the Taylor & Francis Web site at
http://www.taylorandfrancis.com

and the CRC Press Web site at
http://www.crcpress.com

To Lennart, Victoria, Emilia, and Louis

Contents

Foreword

Since their beginnings in the 1930s nonparametric methods have found entrance into various fields of application such as medicine, natural sciences, sociology, psychology, and economics. To a great extent the theoretical foundations of nonparametric methods were completed in the late 1960s, and were described in several monographs. The authors of this foreword published their textbook *Nichtparametrische statistische Methoden* (DeGruyter, 1978) as one of the first German books about these "novel" statistical methods. In those days it was foreseeable that the rapid development of electronic data processing, especially the triumph of PCs in the early 1980s, was of crucial importance for the further success of nonparametric statistics. Thomas Hettmannsperger (1984, p. viii) in his monograph *Statistical Inference Based on Ranks* (Wiley) made the following visionary statement:

> Hopefully, in the 1980s we will see the computer implementation and more widespread use of these efficient and robust statistical methods.

The current textbook by Markus Neuhäuser lies directly within this trend. The focus is on rank-based methods. Ranks, a simple discrete surrogate for the sample variables and asymptotically highly correlated with them, are very important for the construction of efficient and robust tests. In addition to the classical rank tests, the book pays particular attention to the test of Baumgartner, Weiss, and Schindler due to its remarkable power.

The author of this book contributed several research papers about the Baumgartner et al. test as well as other tests. This gives the book a special personal touch. A particular strength is the consequent orientation to computer-based methods. Broad attention is given to permutation tests, exact or approximate, and bootstrap methods, with a coherent implementation in SAS. Supported by a thorough literature review, the different tests are investigated with regard to their power. Here, the reader learns a lot about the actual type I error rate, p-values, robustness, and power, largely illustrated by their own simulation studies.

Numerous real-life data come from various areas, including the Bible, and their analyses provide greatly diversified reading. Noteworthy is the discussion of not so well-known methods such as, for instance, the Cucconi test, the treatment of ties, the combination of p-values, and sample size determination. Moreover, adaptive tests for unknown distributions are discussed.

With the consequent usage of the possibilities provided by modern PCs, the author reveals novel results about nonparametric methods, and presents convincing arguments for their further propagation.

We hope that this book puts nonparametric methods back on the map for a new triumph.

Herbert Büning
Götz Trenkler

Preface

This book deals with nonparametric statistical tests, and has a special focus on a computational approach. Computer-intensive methods such a bootstrap and permutation tests are discussed in detail. Programs are offered to perform the presented tests. An outline of the book can be found in the introductory Chapter 1.

I have benefited from many teachers and colleagues. I wish to thank Professors Herbert Büning and Götz Trenkler not only for their foreword to this book, but also for all that I learned from them during previous years. I also learned a lot from Ludwig Hothorn; I owe him a debt of thanks for his diverse support and his helpfulness since 1994, the year I started as a PhD student at his institute, which is now the Institute of Biostatistics at the Leibniz University of Hannover. I also thank Edgar Brunner for many inspiring discussions on nonparametric topics, and Bryan Manly for providing his program to carry out the D.O. test as well as for initiating the contact with Chapman & Hall/CRC. Furthermore, I wish to acknowledge the support of my colleagues at the Department of Mathematics and Statistics, University of Otago, during the time I worked there. Karl-Heinz Jöckel gave advice and support, in particular within the area of computer-intensive nonparametric methods, during the years I worked at his institute at the University Hospital in Essen. Last but not least, I would like to express my gratitude to Graeme Ruxton for reviewing and proofreading the whole manuscript.

I also thank my colleagues, assistants, and students in Remagen. Dorothee Ball, David Endesfelder, Ann-Kristin Leuchs, Martina Lotter, Denise Rey, and Andreas Schulz contributed to this book; they helped with regard to SAS and R programs, LaTex problems, figures, and the bibliography. They, as well as some other students, also found some errors in the manuscript. Nonetheless, the author should be blamed for all still existing errors. Reports of errors as well as any other comments and suggestions are very welcome and can be sent to neuhaeuser@rheinahrcampus.de.

Finally, I would like to thank Kathrin Mönch, at Oldenbourg, where the German version of this book was published in 2010, and Rob Calver, at Chapman & Hall/CRC for their advice and support.

I hope that this book provides readers with new insights and can help them analyze their data in the most appropriate way.

Markus Neuhäuser   Remagen (Germany)

List of Figures

List of Tables

Chapter 1

Introduction and Overview

"Normality is the exception rather than the norm in applied research."
(Nanna and Sawilowsky, 1998, p. 64).

Nonparametric methods developed into an important area of modern statistics during the second half of the 20th century. Important reasons for the success of nonparametric methods were their broad applicability as well as their high efficiency (Hollander and Wolfe, 1999, p. 13). Nonparametric methods neither require a specific distributional assumption nor a high level of measurement. Several nonparametric tests can be applied in case of nominal or ordinal data. This "universal applicability" is their main advantage according to Büning and Trenkler (1994, p. 2).

In statistical practice, a normality assumption is often not justified. As Büning (1997) and Nanna and Sawilowsky (1998) pointed out, normality is the exception rather than the rule. Micceri (1989) investigated 440 large data sets from psychological research. Regarding symmetry and tails, less than 7% of these data sets were similar to a normal distribution. Every data set was non-normal at the significance level $\alpha = 0.01$. Right-skewed distributions can also often be observed in other areas, such as genetics (Tilquin et al., 2003) and ecology (Mayhew and Pen, 2002, p. 143). These right-skewed distributions can be expected for many variables because there often is a lower bound, such as zero (Gould, 1996, pp. 54–56). The importance of right-skewed distributions for statistical practice was also noted by Büning (2002).

In applications, non-normal data were sometimes transformed in order to normalize them. However, a transformation is often not necessary, because there are powerful nonparametric methods. Moreover, a transformation can be problematic. Piegorsch and Bailer (1997, p. 130) argued that a transformation must be selected based on a priori information; they wrote: "[A] transformation ... must be motivated from previous experimental or scientific evidence. Unless determined a priori, transforms can be misused to inflate or mitigate observed significance in a spurious fashion." Furthermore, the hypotheses before and after the transformation may differ (Games, 1984; McArdle and Anderson, 2004; Wilson, 2007).

According to Rice and Gaines (1989), there is a further problem when a transformation is applied in case of small sample sizes. When a transformation cannot be selected a priori, the alternative is to find a suitable transformation by trial and error. However, the result of such a procedure can be "inappropriate when sample sizes are small ... because residual analysis cannot reliably determine the suitability of a transformation" (Rice and Gaines, 1989, p. 8183). Moreover, a transformation might solve one problem, but at the same time deliver another. For instance, a transformation that eliminates heteroscedasticity tends to skew the transformed data (Rice and Gaines, 1989). Hence, transformations are rarely helpful, especially when sample sizes are small (Neuhäuser, 2010).

A normal distribution is impossible, and cannot be obtained via transformation, when the data are nominal or ordinal. Ordinal scales are common, for instance, when data are gathered with questionnaires (Gregoire and Driver, 1987). Moreover, precise measurements are sometimes not possible, for example in psychological (Sheu, 2002) and biomedical research (Rabbee et al., 2003).

Overview

The focus of this book is on statistical tests. The topics of estimation and confidence intervals are only briefly discussed in Chapter 13.

Chapters 2 to 7 deal with the two-sample problem, one of the most important test problems in statistical applications. For a start, nonparametric two-sample tests for the so-called location-shift model are investigated. Three tests—to be precise, the Fisher-Pitman permutation test, the Wilcoxon rank sum test, and the Baumgartner-Weiss-Schindler test—are presented in detail. Note that the first two tests were recently proposed by Lehmann (2009).

Permutation tests have priority in Chapter 2. Those tests were already introduced in the 1930s, but could be established in applications only recently since fast algorithms and, in particular, powerful PCs became available. The motivation for asymptotic tests is often that they are good approximations for exact methods such as permutation tests (Rodgers, 1999). However, in many situations, exact tests can easily be carried out nowadays. Hence, it is, according to Berger et al. (2008), an irony that asymptotic methods are justified as good approximations, "with no mention of the fact that the gold standard analysis [i.e., the exact test] they are trying to approximate are themselves readily available" (Berger et al., 2008, p. 237). In a Socratic dialogue, also written by Berger (2009), Socrates asked: "If you can observe the exact p-value, then why would you go on to attempt to approximate it?"

From Chapter 3 the assumption of homoscedasticity, that is identical variabilities in the two groups to be compared, is abandoned. The topics of Chapter 3 are location-scale tests, tests for the nonparametric Behrens-Fisher problem, and tests for a difference in variability. Bootstrap tests are also introduced in Chapter 3.

Chapter 4 discusses tests for the general alternative, including the (Kolmogorov-)Smirnov test, before ordered categorical and discrete numerical data are investigated in Chapter 5. The conservatism of permutation tests is debated in Chapter 6. In all chapters, example data sets are considered in order to illustrate the methods. In addition, Chapter 7 presents five further examples for the comparison of two groups. The examples originate from different applications. In addition to an educational experiment and a clinical trial, data sets from ecology and epidemiology are analyzed.

The topics of Chapters 8 and 9 are one-sample tests and tests for more than two groups, respectively. In addition to well-known one-sample tests such as the sign test and Wilcoxon's signed rank test, a modification suggested by Pratt (1959), a permutation test with original observations, and a one-sample bootstrap test are presented. As tests for more than two groups, the following tests are described in detail: the Kruskal-Wallis test, the permutation F test, the Jonckheere-Terpstra trend test, tests for umbrella alternatives, and the Friedman and Page tests for multiple dependent groups.

Chapter 10 treats the concepts of independence and correlation. In addition to the χ^2 test and the likelihood-ratio test, this chapter deals with correlation coefficients and corresponding exact tests. Stratified tests such as the van Elteren test, as well as combination tests, are discussed in Chapter 11. Chapter 12 presents tests for nonstandard situations and some complex designs. The applicability of computer-intensive methods such as bootstrap and permutation tests for nonstandard situations and complex designs is one of the most important advantages of these methods.

In this book, SAS programs (SAS Institute Inc., Cary, North Carolina) are presented to enable the reader to carry out the different statistical tests. R code can be found in the appendix. Furthermore, the appendix provides some basic ideas on the different levels of measurement, statistical tests, and multiple testing. These sections should offer a repetition of key points. Some previous knowledge of statistical tests is required for reading this book. Moreover, some basic knowledge of SAS is required in order to understand the presented SAS programs.

In the literature, the terms *nonparametric* and *distribution-free* are often used synonymously. These terms can be defined as follows (Büning and Trenkler, 1994, p. 1): A distribution-free method is based on a statistic whose distribution does not depend on the specific pattern of the population distribution, that is, the distribution of the underlying data. On the other hand, a method is nonparametric if it does not make any statement about individual parameters of the population distribution. However, as per Büning and Trenkler (1994), we do not rigorously separate the two terms and we mainly use the term *nonparametric*.

Chapter 2

Nonparametric Tests for the Location Problem

As mentioned in the introduction, we first consider the two-sample problem. In this chapter we focus on the location-shift model; thus, we assume that the distributions of the two samples are identical except for a possible difference in location.

The two sample sizes are n_1 and n_2, and let $X_1, \ldots, X_{n_1}$ and $Y_1, \ldots, Y_{n_2}$ denote the two corresponding independent random samples. Let $N = n_1 + n_2$ be the total sample size, and $\bar{X}$ and $\bar{Y}$ the arithmetic means. The observations within each sample are independent and identically distributed with the (cumulative) distribution functions F and G. For a start we assume that F and G are continuous, and in addition that the only possible difference is a shift in location: $F(t) = G(t - \theta)$ for all t, $-\infty < \theta < \infty$.

The null hypothesis is H_0: $\theta = 0$; that is, there is no difference at all between the distribution functions F and G. This null hypothesis can also be expressed as $F = G$ or as $P(X_i < Y_j) = 1/2$ ($i \in \{1, \ldots, n_1\}, j \in \{1, \ldots, n_2\}$), as these expressions are identical in the location-shift model considered here (see, e.g., Horn, 1990). The alternative H_1 states $\theta \neq 0$; thus the distribution functions F and G differ. However, they can differ in location only.

2.1 The Fisher–Pitman Permutation Test

When F and G are normal distributions, the two-sample t test is the uniformly most powerful unbiased test for the location problem H_0 versus H_1. Although this is not the case for other distributions, the t test is often used in statistical practice because it is robust regarding the underlying distributions. For instance, Keller-McNulty and Higgins (1987, p. 18) wrote: "Claims of robustness of the independent samples t statistic have led practitioners to apply this statistic widely ... with little regard for assumptions of normality."

However, the work of Micceri (1989), already cited in the introduction, demonstrated that real distributions often deviate from a normal distribution much more than those used in studies to investigate the robustness. Therefore, Sawilowsky and Blair (1992) investigated the robustness of the t test for some of the distributions identified by Micceri (1989). They found distinct differences between the actual and the nominal type I error rate, and conclude: "The degree of nonrobustness seen in these instances was at times more severe than has been previously reported" (Sawilowsky and Blair, 1992, p. 359). The t test is unreliable, that is, not robust with regard to size, especially in case of extreme skewness.

As a consequence, the t test should not be applied as usual if F and G differ from normal distributions. To be precise, the decision should not be based on the t distribution. An alternative approach is to retain the test statistic, but to use the permutation null distribution for inference. The permutation null distribution can be generated as follows (see, e.g., Good, 2000, or Manly, 2007): At first, all permutations must be generated; that is, one has to list all possible combinations of the N values to the two samples, subject to the restriction that the two sample sizes are the same as in the actual data. Under the null hypothesis there is no difference between the two samples. Thus, each value could also occur with the same probability in the other group. Therefore, any permutation, including the actual data, has the same probability under H_0.

For each permutation the test statistic must be calculated. The next step is to observe how extreme the value of the test statistic, calculated with the original data, is in comparison to all possible values. To be precise, the p-value of the permutation test is the probability of those permutations whose test statistic has an extreme value, that is, a value as or more supportive of the alternative than the value based on the original data. Because all permutations have an identical probability in the model considered here, the permutation null distribution is simply the relative frequency distribution of the test statistic. Thus, the p-value is the proportion of permutations with extreme values. When this p-value is smaller than, or equal to, the nominal significance level α, the null hypothesis can be rejected.

When the test statistic is multiplied by a constant factor, the p-value does not change. Important for the calculation of the p-value is the ordering of the permutations, not the exact value of the test statistic. The latter, of course, is of crucial importance when the test statistic is compared with a t distribution. As a result, the test statistic can be expressed in a simpler form (Manly, 2007, pp. 16–17). For instance, one can use the difference of means $\bar{X} - \bar{Y}$ instead of the t statistic

$$t = \frac{\bar{X} - \bar{Y}}{S \cdot \sqrt{1/n_1 + 1/n_2}},$$

where S is calculated using the values of both groups as follows:

$$S = \sqrt{\frac{1}{N-2} \left(\sum_{i=1}^{n_1}(X_i - \bar{X})^2 + \sum_{j=1}^{n_2}(Y_j - \bar{Y})^2 \right)}.$$

Bradley (1978, pp. 79–80) showed in detail that the t statistic is a monotone function of $\bar{X} - \bar{Y}$. A further equivalent test statistic is the sum of the values of one group, for example $\sum_{i=1}^{n_1} X_i$. This test statistic was proposed by Pitman (1937). In order to receive a one-sided rejection region despite the two-sided alternative H_1, one can use the statistic

$$P = \left| \sum_{i=1}^{n_1} X_i - n_1 \cdot \frac{n_1 \bar{X} + n_2 \bar{Y}}{N} \right|$$

rather than the sum (Pitman, 1937).

The permutation test with one of the above-mentioned equivalent test statistics is called the Fisher-Pitman permutation test (see, e.g., Neuhäuser and Manly, 2004) or *randomization* test (Edgington and Onghena, 2007). It is a nonparametric test (see, e.g., Romano, 1990). Because the test utilizes the numerical values of the observations, at least interval measurement of the variable being studied is required (Siegel, 1956, p. 152).

Because the Fisher-Pitman test is a permutation test based on the t statistic, the question arises as to how powerful this test is in comparison with the usual t test. To answer this question, the concept of relative efficiency is introduced first.

The finite relative efficiency (Büning and Trenkler, 1994, pp. 275–278) of test T_1 to test T_2 is defined as the ratio m/n, where test T_1 has the same power with n observations as test T_2 with m observations, for the same alternative and the same significance level. For the asymptotic relative efficiency, there are different approaches. Here, the concept of Pitman is applied:

Definition of the asymptotic relative efficiency (A.R.E.) according to Pitman (Büning and Trenkler, 1994, p. 279):
Let $[T_{1n}]$ and $[T_{2n}]$ be sequences of test statistics for $H_0 : \theta \in \Omega_0$ versus $H_1 : \theta \notin \Omega_0$ with an identical level α with the corresponding sequences of power functions $[\beta_{1n}]$ and $[\beta_{2n}]$; $[m_i]$ and $[n_i]$ are increasing sequences of positive integers, and with $\lim_{i\to\infty} \theta_i = \theta_0 \in \Omega_0$ we have

$$\lim_{i\to\infty} \beta_{1n_i}(\theta_i) = \lim_{i\to\infty} \beta_{2m_i}(\theta_i) = \beta, \quad 0 < \beta < 1.$$

Then, the A.R.E. of test T_1 to test T_2 is defined by

$$E_{T_1,T_2} = \lim_{i\to\infty} \frac{m_i}{n_i},$$

provided that this limit exists and is the same for all such sequences $[m_i]$ and $[n_i]$. □

This A.R.E. neither depends on α nor on the (asymptotic) power β, in contrast to the finite relative efficiency. Further details and examples of calculations of the A.R.E. can be found in Büning and Trenkler (1994, pp. 280–285).

The asymptotic relative efficiency of the Fisher-Pitman permutation test to the t test is 1 (Lehmann, 2009). Often, and in particular when sample sizes are large, the results of the Fisher-Pitman test and the t test hardly differ.

However, some examples with small sample sizes, considered below in this book, show that the results can be very different.

2.1.1   Example

Let us consider an example data set of Good (2001, p. 56). Cell counts of four cultures treated with vitamin E, and four other cultures that form a control group were observed. The cell counts in the control group were 12, 22, 34, and 95. Within the treated group, here denoted as group 1, the cell counts were 90, 110, 118, and 121. These latter values are larger on average, the mean in group 1 is $\bar{x} = 109.75$. The mean of the control group (group 2) is $\bar{y} = 40.75$, only. Thus, the difference of means is $\bar{x} - \bar{y} = 69$.

The null hypothesis states that there is no difference between the two groups of cell cultures—that is, all eight values come from the same underlying distribution. Hence, each of the values could have been occurred with the same probability in the other group as well. Thus, we generate all possible combinations for how eight values can be divided into two groups of size 4. These permutations all have an identical probability under the null hypothesis. In total, there are $\binom{N}{n_1} = \binom{8}{4} = 70$ permutations. Consequently, each permutation has the probability $1/70 = 0.0143$.

The test statistic, here the difference of means, has its largest value if the four largest observations (95, 110, 118, 121) were assigned to group 1, and if the four smallest observations (12, 22, 34, 90) were in group 2. In this case the difference of means is 71.5. The second-largest difference of means occurs for the actual data. The third-largest value of the difference is 61.5 (see Table 2.1), one gets this value if the observations 90, 95, 118, and 121 are assigned to group 1. For this permutation the value of the test statistic, that is, the difference of means, is smaller than for the actual observations. Thus, the probability $P_0(\bar{X} - \bar{Y} \geq 69)$ that the difference $\bar{X} - \bar{Y}$ is at least as large as 69, and the difference for the actual data is $2/70 = 0.0286$ according to the permutation null distribution. The reason is that exactly two permutations give a difference of 69 or larger. The subscript 0 in the above probability indicates that this probability is calculated under the assumption that the null hypothesis is true.

In order to determine the p-value of the (two-sided) permutation test we have to consider all permutations whose evidence against the null hypothesis is at least as large as for the observed permutation. Hence, permutations with a large negative difference between means must also be considered, because a negative difference with a large absolute value gives as much evidence against the null hypothesis, and for a difference between the two groups, as a positive difference with the same absolute value. Thus, the probability $P_0(|\bar{X} - \bar{Y}| \geq 69)$ is the p-value of the two-sided test. This probability is $4/70 = 0.0571$ because the absolute value of the difference is 69 or larger for

TABLE 2.1: The exact permutation null distribution of $\bar{X} - \bar{Y}$ for the example of Good (2001)

Possible Value of $\bar{X} - \bar{Y}$	Probability (= proportion within the 70 permutations)
-71.5	$1/70$
-69	$1/70$
-61.5	$1/70$
-57.5	$1/70$
-56	$1/70$
...	...
56	$1/70$
57.5	$1/70$
61.5	$1/70$
69	$1/70$
71.5	$1/70$

four permutations (see Table 2.1). As a result, the difference between the two groups is not significant at the significance level $\alpha = 5\%$.

An identical p-value of $4/70 = 0.0571$ results when the test statistic

$$P = \left| \sum_{i=1}^{n_1} X_i - n_1 \cdot \frac{n_1 \bar{X} + n_2 \bar{Y}}{N} \right|$$

is used. Table 2.2 displays the exact permutation null distribution of P. For the actual data we have $P = 138$. According to Table 2.,2 the probability for $P \geq 138$ is $4/70$ because we have $P = 138$ for two permutations and $P = 143$ for additional two permutations. All other 66 permutations have smaller values of the test statistic P.

It should be mentioned that there are in total $8! = 40,320$ permutations of $N = 8$ values. However, many of these permutations only rearrange the values within groups. Such rearrangements within groups are irrelevant because they do not have any effect on the test statistic. As a consequence, it is sufficient to confine the attention to the $\binom{N}{n_1}$ "permutations" that differently partition the N observations to two groups of sizes n_1 and n_2 (Bradley, 1968, pp. 78–79).

The permutation null distribution depends on the observed data. Therefore, the permutation test is a conditional test, given the observed values of the two samples. Because of this dependency on the observations, the permutation null distribution cannot be tabulated, and the test requires a large amount of computation. That is the reason that the test was hardly used in practice although it was proposed in the 1930s (Fisher, 1936; Pitman, 1937), and its theoretical key benefits have been known for several decades (e.g., Lehmann and Stein, 1949; Hoeffding, 1952). For instance, Bradley (1968, p. 84) wrote that the permutation test is "almost never quick ... seldom practical, and of-

TABLE 2.2: The exact permutation null distribution of P for the example of Good (2001)

Possible Value of P	Probability (= proportion within the 70 permutations)
6	2/70
14	2/70
16	2/70
17	2/70
24	2/70
27	2/70
28	2/70
29	2/70
32	2/70
34	2/70
36	2/70
37	2/70
39	4/70
40	2/70
42	2/70
44	2/70
45	2/70
47	2/70
50	2/70
51	2/70
54	2/70
55	2/70
56	2/70
59	2/70
60	2/70
62	2/70
67	2/70
70	2/70
82	2/70
112	2/70
115	2/70
123	2/70
138	2/70
143	2/70

ten ... not even feasible." According to May and Hunter (1993, p. 402), the Fisher-Pitman test remained "in relative obscurity." However, nowadays fast algorithms and powerful PCs are available; hence, the test has been recommended several times (see, e.g., Crowley, 1992; Gebhard, 1995; Thomas and Poulin, 1997; Berry et al., 2002). The Fisher-Pitman permutation test is implemented, for example, in R (Hothorn and Hornig, 2002) and implemented in the statistical software package StatXact (Cytel Software Corporation, Cambridge, Massachusetts); the latter software calls this test "permutation with general scores test."

2.1.2 Implementation in SAS

In SAS the Fisher-Pitman permutation test can be carried out using the procedure NPAR1WAY. To be precise, within the PROC NPAR1WAY statement, the option SCORES=DATA is needed. Because of this option, the Fisher-Pitman test based on the original data is applied rather than a rank test. Moreover, the EXACT statement is necessary to perform an exact, that is a permutation test. In order to analyze the example data of Good (2001) with four observations per group, the following SAS program can be used:

```
DATA example1;
   INPUT group count;
CARDS;
1 90
1 110
1 118
1 121
2 12
2 22
2 34
2 95
;

PROC NPAR1WAY SCORES=DATA;
   CLASS group;
   VAR count;
   EXACT;
RUN;
```

The output presents the result of the exact permutation test, along with other things. A one-sided (see Section 2.5) as well as the two-sided p-value are given:

```
Exact Test
 One-Sided Pr >=  S            0.0286
 Two-Sided Pr >= |S - Mean|    0.0571
```

2.1.3   Approximate Permutation Test

When all permutations are considered, the permutation test is also called the exact permutation test. However, a permutation test can also be carried out as an approximate test based on a simple random sample out of all possible permutations (Edgington and Onghena, 2007, Section 3.6; Good, 2000, Section 13.2). This approach is appropriate in case of large sample sizes, as the number of permutations can then be huge. When an approximate permutation test is applied, the original data must be included within the random sample of permutations. Thus, one needs a simple random sample of $M - 1$ (non-observed) permutations, and the observed permutation is added to this sample. When the test is performed in this way, the p-value cannot be smaller than $1/M$. As a consequence, the probability to receive a p-value smaller than or equal to α is, under the null hypothesis, not larger than α (Edgington and Onghena, 2007, p. 41). This is necessary to maintain the nominal significance level. Nowadays, values such as $M = 10,000$ are standard. With such a large value for M, there obviously cannot be a large difference, whether or not the observed permutation is added. The p-value can change in the fourth digit by 0.0001 only.

An approximate Fisher-Pitman permutation test can be carried out with the SAS procedure NPAR1WAY, when the option MC, for "Monte Carlo estimation of exact p-values," is added in the EXACT statement:

```
PROC NPAR1WAY SCORES=DATA;
  CLASS group;
  VAR count;
  EXACT / MC;
RUN;
```

The relevant part of the output then displays the following details:

```
Monte Carlo Estimates for the Exact Test
  One-Sided Pr >= S
  Estimate                      0.0270
  99% Lower Conf Limit          0.0228
  99% Upper Conf Limit          0.0312

  Two-Sided Pr >= |S - Mean|
  Estimate                      0.0552
  99% Lower Conf Limit          0.0493
  99% Upper Conf Limit          0.0611

  Number of Samples             10000
  Initial Seed                  940250001
```

In addition to the estimation of p-values, confidence intervals are given.

With a sample of 10,000 permutations, one can apply the central limit theorem to obtain a confidence interval. Let $\hat{p}$ denote the estimated p-value; then

$$\hat{p} \pm z_{1-\alpha/2} \cdot \sqrt{\frac{\hat{p}(1-\hat{p})}{10\,000}}$$

is a $(1-\alpha)$ confidence interval for the p-value, where $z_{1-\alpha/2}$ is the $(1-\alpha/2)$ quantile of the standard normal distribution.

The confidence level $1-\alpha$ can be changed using the following statement:
`EXACT / MC ALPHA=0.05;`

Moreover, the "Number of samples," that is, the number of permutations used to estimate the p-value, can be specified in the SAS program:
`EXACT / MC N=1000;`

The "Initial Seed" for generating the random sample of permutations can be specified as follows:
`EXACT / MC SEED=3579;`

The three options ALPHA=, N=, and SEED= mentioned above make sense only if an approximate permutation test is performed. Therefore, if at least one of these options is specified, a Monte Carlo estimation of the p-value is carried out, even when MC is not added to the program. Thus, the following statements are equivalent:
`EXACT / MC SEED=3579;`
`EXACT / SEED=3579;`

One further option in relation to the EXACT statement is MAXTIME = value. With this option, the time (in seconds) can be specified that SAS can use at most in order to compute the p-value of a permutation test. When more time is needed, the computation stops. This option is available for both an exact as well as an approximate calculation of the p-value. If an exact p-value is computed (i.e., no MC option), the computation stops when too much time is needed without automatically changing to the approximate test.

By default, SAS uses 10,000 permutations for the approximate permutation test. This standard value was used to produce the above output, although in the example there are seventy permutations only. Hence, at least some permutations are used several times, the permutations are drawn "with replacement." In situations in which an approximate test is applied in practice, the sample sizes are much larger, and therefore there are many more than 10,000 permutations. Nevertheless, it is still possible—although not very likely—that some permutations are used more than once with the above-mentioned program using the NPAR1WAY procedure.

Such a draw "with replacement" of the permutations decreases the power of the approximate permutation test (Opdyke, 2003). Opdyke (2003) presented an SAS program in order to carry out a permutation test where the permutations are drawn "without replacement." We only refer to this program because the gain in power is small when the repetition of permutations is avoided. According to Opdyke (2003, p. 40), the tests are of "practical equivalence." The reason that the difference is small in statistical practice is as follows: When sample sizes are small, an exact permutation test should be carried out. If the sample sizes are so large that an approximate test is selected, the number of possible permutations is huge. Therefore, the probability of drawing any permutation more than once is extremely small, even when several thousand permutations are drawn (Opdyke, 2003).

Please note that the difference between with and without replacement applies to the draw of permutations only. The observed values will be drawn without replacement to form the permutations. This is the case for any permutation test, although it is different for the bootstrap test discussed in Section 3.3.

If the sample sizes are "large," the kurtosis is "small" and $1/5 \leq n_1/n_2 \leq 5$, the permutation null distribution can be approximated by the t distribution with $n_1 + n_2 - 2$ degrees of freedom (Siegel, 1956, pp. 154–155). The possibility of this approximation also demonstrates that the results of the Fisher-Pitman permutation test and the t test can be very similar. Further alternatives approximate the permutation null distribution based on its moments (see, e.g., Box and Andersen, 1955) or model the tails of the distribution with a generalized Pareto distribution (Knijnenburg et al., 2009).

Hereafter, the Fisher-Pitman permutation test is abbreviated as FPP test.

2.2   The Wilcoxon Rank Sum Test

The Wilcoxon rank sum test, which is equivalent to the Mann-Whitney U test, is the most popular nonparametric alternative to the t test in statistical practice (van den Brink and van den Brink, 1989). Instead of computing the sum $\sum_{i=1}^{n_1} X_i$, the test statistic of the Wilcoxon test is the sum of the corresponding ranks. The rank of an observation is 1 plus the number of observations from both samples that are smaller than this observation. The rank sum can also be expressed as a linear rank statistic.

Definition of a linear rank statistic (see, e.g., Büning and Trenkler, 1994, p. 127):

A statistic of the form $T = \sum_{i=1}^{N} g(i)V_i$, where $g(i)$, $i = 1, \ldots, N$, are arbitrary scores, is called a linear rank statistic for the two-sample problem. We have $V_i = 1$, if the i-th smallest of the N values comes from group 1, otherwise $V_i = 0$. □

Under the null hypothesis H_0, expectation and variance of a linear rank statistic T can be expressed as follows:

$$E_0(T) = \frac{n_1}{N} \sum_{i=1}^{N} g(i) \quad \text{and}$$

$$\text{Var}_0(T) = \frac{n_1 n_2}{N^2(N-1)} \left[N \sum_{i=1}^{N} g^2(i) - \left(\sum_{i=1}^{N} g(i) \right)^2 \right].$$

In addition, the standardized linear rank statistic $\frac{T - E_0(T)}{\sqrt{\text{Var}_0(T)}}$ is, under H_0, for large samples ($n_1, n_2 \to \infty$ with $n_1/n_2 \to \lambda \neq 0, \infty$) asymptotically standard normal (see, e.g., Büning and Trenkler, 1994, p. 130). The scores of the Wilcoxon test are $g(i) = i$. Hence, the test statistic is the rank sum of group 1: $W = \sum_{i=1}^{N} i \cdot V_i$. Under H_0, we have $E_0(W) = n_1(N+1)/2$ and $\text{Var}_0(W) = n_1 n_2 (N+1)/12$.

In contrast to the FPP test, a test based on a linear rank statistic can also be applied to ordinal data. The Wilcoxon test was recommended for this situation (e.g., Rahlfs and Zimmermann, 1993; Nanna and Sawilowsky, 1998).

The Wilcoxon rank sum test is the locally most powerful rank test for a logistic distribution (see, e.g., Janssen, 1998, p. 27). A locally optimal rank test maximizes the power "in the proximity of the null hypothesis;" formally it is defined as follows:

Definition of a locally most powerful rank test (see, e.g., Randles and Wolfe, 1979, p. 295):
For a specified distribution F and the test problem H_0: $\theta = 0$ versus $H_1^>$: $\theta > 0$, a rank test is called locally most powerful if there exists some $\varepsilon > 0$ such that this test is the uniformly most powerful rank test for $0 < \theta < \varepsilon$ and any possible significance level. $\qquad\square$

The logistic distribution has the density function

$$f(x) = \frac{g \exp(-gx)}{(1 + \exp(-gx))^2}, \quad -\infty < x < \infty,$$

where g is a constant. This distribution is symmetric. With the choice $g = \frac{\pi}{\sqrt{3}}$, the standardized logistic distribution results, that is, a distribution with expected value 0 and variance 1 (Malik, 1985). The shape of this distribution is very similar to a standard normal one. Therefore, the Wilcoxon rank sum test has a relatively high power for normally distributed data and other symmetrical distributions with medium to large tails (Büning and Trenkler, 1994, p. 304).

The asymptotic relative efficiency (A.R.E.) of the Wilcoxon rank sum test to the t test has a lower limit of 0.864; an upper limit does not exist. For an exponential distribution, the A.R.E. is 3.0; for a normal distribution, $3/\pi$

= 0.955 (Hodges and Lehmann, 1956; Lehmann, 2009). The high efficiency of the Wilcoxon test is therefore astonishing, as the ranks only, and not the complete information, is used for the test. Furthermore, the simplicity of the scores $g(i) = i$ probably also contributes to the popularity of the Wilcoxon test.

The Wilcoxon rank sum test can be performed based on the asymptotic normality of the test statistic or on the permutation null distribution of W. For the asymptotic test, the standardized statistic

$$Z_W = \frac{W - \frac{n_1(N+1)}{2}}{\sqrt{\frac{n_1 n_2 (N+1)}{12}}}$$

is calculated, which is asymptotically standard normal (see above). Hence, the null hypothesis can be rejected in a (two-sided) asymptotic test at level α, if $|Z_W| \geq z_{1-\alpha/2}$; the p-value is $2(1 - \Phi(|Z_W|))$. Here, $z_{1-\alpha/2}$ denotes again the $(1 - \alpha/2)$ quantile, and Φ is the distribution function of the standard normal distribution.

In contrast to the FPP test, the permutation null distribution can be tabulated when the Wilcoxon test is carried out as a permutation test. The reason is that all values needed to calculate the test statistic (i.e., the ranks from 1 to N) are known in advance. The experiment only provides information on how the ranks are subdivided into the two groups. Please recall that we assumed continuous distribution functions F and G up to now.

Which of the two ways of performing the test is more appropriate? In the *population model* (Lehmann, 2006, Chapter 2), the inference is referred to a defined population that has been randomly sampled. Using this classical model-based inference, the asymptotic distribution can be used if the sample size is not too small. However, what does "not too small" exactly mean? There are different rules of thumb. Brunner and Munzel (2002, p. 63) wrote that the normal approximation is acceptable if $\min(n_1, n_2) \geq 7$, as long as no ties occur, that is, when all observed values differ. Büning and Trenkler (1994, p. 134) suggested that n_1 or $n_2 > 25$ is needed. In practice, however, asymptotic p-values are sometimes given for much smaller sample sizes (Mundry and Fischer, 1998). An example is an asymptotic test for $n_1 = 12$ and $n_2 = 3$, published in 2002 in *Nature* (Blomqvist et al., 2002). In such a case, the exact permutation distribution should be used to determine the p-value.

The *randomization model* of inference does not require that populations have been randomly sampled, only that the groups or treatments have been assigned to the experimental units at random (Lehmann, 2006, p. 5; see also Ludbrook and Dudley, 1994). In this design-based framework, the p-value should be that from the permutation distribution.

Ludbrook and Dudley (1998) surveyed more than 250 prospective, comparative studies reported in five reputable biomedical journals. They found that randomization rather than random sampling is the norm in biomedical research (Ludbrook and Dudley, 1998, p. 127). Only 4% of the studies used random sampling of defined populations to construct experimental groups. Furthermore, the sample sizes were often small. The median group sizes (per journal) varied from four to nine in the case of random sampling and from six to twelve in randomized studies. This suggests that exact permutation tests should be used for the vast majority of biomedical studies (see also Neuhäuser, 2005a). This is probably also true for many other research studies, for example, those in psychology (Hunter and May, 1993).

In case of a significant result within the randomization model, the result of the experiment is very unlikely under H_0, and one can conclude that this result is unlikely to have occurred by chance (see, e.g., Ludbrook and Dudley, 1994). But there is no population, to which the inference can be extended. This might be considered a disadvantage of a permutation test. Nevertheless, Berger (2000) argues that this is, in the context of randomized clinical trials, a weakness not of the permutation test, but of the study design.

In randomized clinical trials there usually is no random sample; instead, the included patients are a "convenience sample." Moreover, the representativeness of the patients within a clinical trials is also limited because there are inclusion and exclusion criteria such as the willingness to be randomized, and a run-in selection (Leber and Davis, 1998). In addition, the fact that patients are aware of being observed can have a large effect (Fisher and van Belle, 1993, p. 18). Thus, external validity and therewith the possibility to generalize the results might be limited and cannot be guaranteed. However, any generalization is excluded if there was no internal validity. According to Berger (2000, p. 1321) "results *may* be generalizable, provided there is internal validity. ... PTs [permutation tests] are a prerequisite for internal validity ... By ensuring internal validity, the PT actually enhances the ability to extrapolate results."

As mentioned above, the Wilcoxon rank sum test is equivalent to the Mann-Whitney U test. The idea of the Mann-Whitney U test is to count how many X_i values follow Y_j values in the combined ordered sample of all N values. Let us consider again the example of Good (2001): $X_1 = 90$, $X_2 = 110$, $X_3 = 118$, $X_4 = 121$, $Y_1 = 12$, $Y_2 = 22$, $Y_3 = 34$, and $Y_4 = 95$. The combined ordered sample therefore is $12, 22, 34, 90, 95, 110, 118, 121$; thus, the order of X and Y values is $yyyxyxxx$. All four X-values succeed the first three Y-values, and three X-values succeed the largest Y-value. Thus, $U = 4 + 4 + 4 + 3 = 15$ results as the Mann-Whitney U statistic.

Formally, the Mann-Whitney U statistic can be defined as follows:

$$U = \sum_{i=1}^{n_1} \sum_{j=1}^{n_2} \phi(Y_j, X_i) \quad \text{with} \quad \phi(a,b) = \begin{cases} 1 & \text{if} \quad a < b \\ 0 & \text{if} \quad a > b. \end{cases}$$

In the presence of ties (see Section 2.7), ϕ can be set to $1/2$.

Based on the value $U = 15$ of the Mann-Whitney statistic, the rank sum W can be calculated as

$$W = U + \frac{n_1}{2}(n_1 + 1).$$

Hence, in the example the rank sum is $15 + 10 = 25$. Obviously, this rank sum of the X-values can be determined directly. The ranks of the X-values are 4, 6, 7, and 8, the ranks of the Y-values are 1, 2, 3, and 5. Thus, the rank sum of the X values, that is, the observations in group 1, is $4 + 6 + 7 + 8 = 25$.

The rank sum test was introduced by Frank Wilcoxon in 1945; Henry Mann and Donald Whitney published their proposal in 1947. The test, however, is older. It was introduced at least six times in addition to Wilcoxon (1945) and Mann and Whitney (1947), see Kruskal (1957). The first who proposed the test was Gustav Deuchler in 1914, at that time at the University of Tübingen, Germany. Deuchler (1914) suggested the following approach: Consider all $n_1 n_2$ pairs (X_i, Y_j). Each pair gets a score, this score being $+1$, -1, or 0, depending on whether the X value is larger, smaller, or equal to the Y-value of the pair. Deuchler's test statistic then is

$$r = \frac{\text{Sum of scores}}{\text{Number of scores with the values } + 1 \text{ or } - 1}.$$

If there are no pairs with $X_i = Y_j$, it follows (Kruskal, 1957) that

$$r = 1 - \frac{2}{n_1 n_2} U.$$

Here we consider the rank sum W as the test statistic. In the example of Good (2001) we have $W = 25$. In order to carry out a permutation test, we have to calculate this rank sum W for all seventy possible permutations. The resulting permutation null distribution is displayed in Table 2.3.

The value of the test statistic is 25 for the observed data. The larger possible rank sum 26 provides stronger evidence against the null hypothesis and for a difference between the two groups. Under the null hypothesis of no difference, the expected value of the rank sum is $n_1(N + 1)/2 = 18$. Because 11 differs from 18 as much as 25, the rank sum 11 gives as much evidence for the (two-sided) alternative as the rank sum 25. And the smallest possible rank sum 10 gives as much evidence for the alternative as the largest rank sum 26. Altogether we have four permutations (with rank sums 10, 11, 25, 26) that must be considered for calculating the p-value. These four permutations give at least as much evidence against H_0 as the observed rank sum 25. Thus, the p-value of the exact Wilcoxon test is $4/70 = 0.0571$.

2.2.1   Implementation in SAS

The Wilcoxon rank sum test can also be carried out with the SAS procedure NPAR1WAY. Instead of SCORES=DATA, the option WILCOXON is needed in the PROC NPAR1WAY statement. The statement EXACT as well as the options regarding this statement can be applied as discussed in Section 2.1. Thus, the following program can be used:

TABLE 2.3: The exact permutation null distribution of the rank sum of the X values for $n_1 = n_2 = 4$

Possible Value of W	Probability (= proportion within the 70 permutations)
10	1/70
11	1/70
12	2/70
13	3/70
14	5/70
15	5/70
16	7/70
17	7/70
18	8/70
19	7/70
20	7/70
21	5/70
22	5/70
23	3/70
24	2/70
25	1/70
26	1/70

```
PROC NPAR1WAY WILCOXON;
  CLASS group;
  VAR count;
  EXACT;
RUN;
```

The following output results:

```
Wilcoxon Two-Sample Test


Statistic (S)              25.0000

Normal Approximation
Z                          1.8764
One-Sided Pr > Z           0.0303
Two-Sided Pr > |Z|         0.0606

t Approximation
One-Sided Pr > Z           0.0514
Two-Sided Pr > |Z|         0.1027
```

```
Exact Test
One-Sided Pr >= S              0.0286
Two-Sided Pr >= |S - Mean|   0.0571
```

```
Z includes a continuity correction of 0.5.
```

```
Kruskal-Wallis Test
  Chi-Square        4.0833
  DF                1
  Pr > Chi-Square   0.0433
```

The rank sum W is denoted by S in this output, Z is the standardized rank sum. In the example we have $Z = (25 - 18)/\sqrt{12} = 2.0207$. However, the SAS output says $Z = 1.8764$. The reason is that SAS applies by default a continuity correction.

The distribution of W is discrete; however, the asymptotic test utilizes a continuous distribution such as the normal distribution. In the SAS, output approximations with the normal and a t distribution were given. Because a discrete distribution is approximated with a continuous one, some authors suggest applying a continuity correction if the sample sizes are small. However, a permutation test is indicated in the case of small samples (Bergmann et al., 2000); therefore, the need for a continuity correction is omitted. For large samples a continuity correction is not necessary because the distribution of W is then less discrete, and a continuity correction can hardly influence the result. Hence, we do not explain the continuity correction here. When applying the procedure NPAR1WAY one can specify the option CORRECT=NO in order to avoid the continuity correction:

PROC NPAR1WAY WILCOXON CORRECT=NO;

With this statement $Z = 2.0207$ results, instead of $Z = 1.8764$. The exact permutation test does not change, it does not depend on any continuity correction.

When using statistical software to perform the Wilcoxon test it is of crucial importance to pay attention to the way that the test is carried out. There can be large differences depending whether the test is performed as an exact or an asymptotic test, or if the latter, whether a continuity correction is applied or not (Bergmann et al., 2000).

The asymptotic test without a continuity correction can be found at the bottom of the SAS output under the heading "Kruskal-Wallis Test." The value of Z^2 is given there as "Chi-Square." The corresponding p-value is $P_0(Z^2 \geq 4.0833) = P_0(|Z| \geq 2.0207) = 0.0433$. It is evident that this asymptotic test should not be applied for this example with $n_1 = n_2 = 4$. However, the asymptotic result is displayed within the SAS output and given here for illustrative purposes.

Although a large variety of linear rank statistics exist, only the Wilcoxon test has been described so far. This test is the most often applied nonparametric test for the location problem (Büning and Trenkler, 1994, p. 135). Hence, it is implemented in most statistical software systems. The Wilcoxon test, equivalent to the Mann-Whitney test, is abbreviated as the WMW test below. Other linear rank statistics are discussed in Section 2.6.

2.3 The Test of Baumgartner, Weiss, and Schindler

In 1998, Baumgartner, Weiss, and Schindler (1998) introduced a new nonparametric test statistic. This statistic is based on ranks, as the WMW statistic, and is defined as follows:

$$B = \frac{1}{2} \cdot (B_X + B_Y), \quad \text{with}$$

$$B_X = \frac{1}{n_1} \sum_{i=1}^{n_1} \frac{\left(R_i - \frac{N}{n_1} \cdot i\right)^2}{\frac{i}{n_1+1} \cdot \left(1 - \frac{i}{n_1+1}\right) \cdot \frac{n_2 N}{n_1}} \quad \text{and}$$

$$B_Y = \frac{1}{n_2} \sum_{j=1}^{n_2} \frac{\left(H_j - \frac{N}{n_2} \cdot j\right)^2}{\frac{j}{n_2+1} \cdot \left(1 - \frac{j}{n_2+1}\right) \cdot \frac{n_1 N}{n_2}}.$$

Here, $R_1 < \cdots < R_{n_1}$ ($H_1 < \cdots < H_{n_2}$) denote the ordered ranks of group 1 (group 2). As for the WMW test, the observations of both groups are pooled in order to determine these combined-samples ranks. Thus, the ranks $R_1, \ldots, R_{n_1}, H_1, \ldots, H_{n_2}$ are a permutation of the integers $1, 2, \ldots, N$. Large values of the test statistic B give evidence against the null hypothesis.

As for the (Kolomogorow-)Smirnov-Test (see Chapter 4), the test of Baumgartner et al. (1998) also makes use of the difference between the empirical distribution functions $\hat{F}$ and $\hat{G}$. To be precise, the term

$$\frac{n_1 n_2}{N} \cdot \int_0^1 \frac{1}{z(1-z)} \cdot \left(\hat{F}(z) - \hat{G}(z)\right)^2 dz$$

is approximated using ranks. The weights $1/(z(1-z))$ emphasize the tails of the distributions (Baumgartner et al., 1998).

Baumgartner et al. (1998) derived the asymptotic distribution of B; they showed that

$$\lim_{n_1,n_2 \to \infty} P_0(B < b) =$$

$$\sqrt{\frac{\pi}{2}} \frac{1}{b} \sum_{i=0}^{\infty} \binom{-1/2}{i} (4i + 1) \int_0^1 \frac{1}{\sqrt{r^3(1-r)}} \cdot \exp\left(\frac{rb}{8} - \frac{\pi^2(4i+1)^2}{8rb}\right) dr$$

$$\text{with } \binom{-1/2}{i} = \frac{(-1)^i \cdot \Gamma\left(i + \frac{1}{2}\right)}{\Gamma\left(\frac{1}{2}\right) \cdot i!}.$$

The convergence of this progression is quite fast, a summation up to $i = 3$ is sufficient (Baumgartner et al., 1998). Please note that Baumgartner et al. (1998) considered the asymptotic test only. Table 2.4 displays the asymptotic distribution of B. Baumgartner et al. (1998) presented the probability $P_0(B < b)$ for just six different values of b. For statistical practice a more detailed table is required, in particular when adjusting the significance level in multiple test problems (see appendix).

In the following, we denote the test with the statistic B as the BWS test. In a simulation study, Baumgartner et al. (1998) compared the asymptotic BWS test with other nonparametric tests. In the location-shift model, however, they restricted the simulations to normal distributions, and demonstrated that the powers of the asymptotic tests BWS and WMW are very similar in this case.

We do not reproduce the simulation results of Baumgartner et al. (1998). The power of the asymptotic BWS test is presented here in Table 2.5 only. In this table the power is given for different t distributions in order to show that the power increases with growing tails. The t distributions were standardized, that is, the variance is 1 in both groups. The power increases with decreasing degrees of freedom (df); thus, the power increases with growing tails. The power of the WMW test also increases with rising tails. However, this increase is stronger for the BWS test; hence, that test is advantageous in the case of few degrees of freedom.

The simulation results presented in Table 2.5 and below are based on 10,000 simulated data sets for any configuration, that is for any estimated power or size. The simulations were performed using SAS, partly Proc-StatXact (Cytel Software Corporation, Cambridge, Massachusetts) was additionally used. When using 10,000 simulation runs the central limit theorem can be used to compute a 95% confidence interval. Let us denote R as the power and $\hat{R}$ as its estimation, then

$$\hat{R} \pm 1.96 \cdot \sqrt{\frac{R(1-R)}{10\,000}}$$

is a 95% confidence interval for the power, or for the actual type I error rate, respectively (see, e.g., Berry, 1995a).

The possible range of R is the interval from 0 to 1. Within this range

TABLE 2.4: Table of the asymptotic distribution of the BWS statistic B

b	$P_0(B \geq b)$	b	$P_0(B \geq b)$	b	$P_0(B \geq b)$
0.1	>0.9999	3.1	0.0244	6.1	0.00087
0.2	0.9904	3.2	0.0217	6.2	0.00078
0.3	0.9382	3.3	0.0193	6.3	0.00070
0.4	0.8487	3.4	0.0172	6.4	0.00063
0.5	0.7468	3.5	0.0154	6.5	0.00056
0.6	0.6480	3.6	0.0137	6.6	0.00051
0.7	0.5588	3.7	0.0122	6.7	0.00046
0.8	0.4810	3.8	0.0109	6.8	0.00041
0.9	0.4142	3.9	0.0098	6.9	0.00037
1.0	0.3573	4.0	0.0087	7.0	0.00033
1.1	0.3088	4.1	0.0078	7.1	0.00030
1.2	0.2675	4.2	0.0070	7.2	0.00027
1.3	0.2323	4.3	0.0062	7.3	0.00024
1.4	0.2023	4.4	0.0056	7.4	0.00022
1.5	0.1765	4.5	0.0050	7.5	0.00019
1.6	0.1543	4.6	0.0045	7.6	0.00017
1.7	0.1352	4.7	0.0040	7.7	0.00016
1.8	0.1186	4.8	0.0036	7.8	0.00014
1.9	0.1043	4.9	0.0032	7.9	0.00013
2.0	0.0918	5.0	0.0029	8.0	0.00011
2.1	0.0810	5.1	0.0026	8.1	0.00010
2.2	0.0715	5.2	0.0023	8.2	0.00009
2.3	0.0632	5.3	0.0021	8.3	0.00008
2.4	0.0559	5.4	0.0019	8.4	0.00007
2.493	0.0500	5.5	0.0017	8.5	0.00007
2.6	0.0439	5.6	0.0015	8.6	0.00006
2.7	0.0390	5.7	0.0013	8.7	0.00005
2.8	0.0346	5.8	0.0012	8.8	0.00005
2.9	0.0308	5.9	0.0011	8.9	0.00004
3.0	0.0274	6.0	0.0010	9.0	0.00004

TABLE 2.5: The simulated power of the asymptotic BWS test as well as the difference in power between the asymptotic tests BWS and WMW for different standardized t distributions

df	Power of the asy. BWS Test	Difference in Power (asy. BWS test – asy. WMW test)
3	0.79	0.016
4	0.69	0.010
6	0.61	0.006
8	0.59	−0.003
16	0.56	−0.005
100	0.54	−0.007

the function $R(1 - R)$ has its maximum at $R = 0.5$. Hence, the confidence interval has its maximum length when $R = 0.5$. In this case we have $1.96 \cdot \sqrt{(0.5(1 - 0.5))/10,000} = 0.0098$. Such a small difference of less than 1% might be negligible for a power estimation. For $R = 0.05$, more appropriate than 0.5 for the estimation of the actual significance level in case of $\alpha = 0.05$, the 95% confidence interval is narrower with $1.96 \cdot \sqrt{(0.05(1 - 0.05))/10\,000} = 0.0043$.

Baumgartner et al. (1998) recommended their (asymptotic) test even for small sample sizes. For instance, they applied the test in an example with $n_1 = n_2 = 5$. However, the asymptotic test can be anticonservative in case of small sample sizes (Neuhäuser, 2000). The actual level is 0.055 for $n_1 = n_2 = 10$ and a nominal level of $\alpha = 0.05$. One may argue that this deviation is small and therefore acceptable. On the other hand, the actual level of a permutation test based on the statistic B hardly differs from the nominal level α (See section 2.4). Hence, the exact permutation test with B might be a good option for small sample sizes. It was suggested by Neuhäuser (2000); results regarding size and power are presented in Section 2.4.

For the example data of Good (2001), the ranks of the X-values are 4, 6, 7, and 8, and the ranks of the Y-values are 1, 2, 3, and 5. Thus, we get $B_X = 1.4323$, $B_Y = 3.6458$, and therefore $B = 2.5391$. The p-value is therefore the probability $P_0(B \geq 2.5391)$. We have to consider the right tail only of the distribution of B as B does not depend on the direction of a possible deviation from H_0, due to the squares in the numerators of B_X and B_Y. The larger the deviation from H_0, the larger the statistic B.

When using the asymptotic distribution of B, the probability $P_0(B \geq 2.5391)$ is smaller than 0.05 (see Table 2.4). However, the sample size is four per group and hence a permutation test should be carried out. The permutation distribution of B is given in Table 2.6. Based on these results we get $P_0(B \geq 2.5391) = 4/70 = 0.0571$. With the example data of Good (2001), the

TABLE 2.6: The exact permutation null distribution of the BWS statistic B for $n_1 = n_2 = 4$

Possible Value of B	Probability (= proportion within the 70 permutations)
≤ 0.75	40/70
0.8789	4/70
0.9440	4/70
0.9766	4/70
1.2370	2/70
1.4974	4/70
1.6276	4/70
1.8555	2/70
1.9206	2/70
2.5391	2/70
3.7109	2/70

p-values of all three permutation tests FPP, WMW, and BWS are identical. This is possible, but it would be quite unlikely in case of larger sample sizes.

2.3.1 Implementation in SAS

There is no ready-to-use SAS procedure in order to perform the BWS test. However, the following SAS macro (according to Neuhäuser et al., 2009) carries out the BWS test as a permutation test using all possible permutations:

```
%MACRO Permtest(indata);
proc iml;

/* Reading the data */
USE &indata;
READ ALL INTO currdata;

/* Computation of ranks */
ranks=RANKTIE(currdata[ ,2]);

/* Calculation of the sample sizes per group */
N_total=Nrow(currdata[ ,2]);
n2=currdata[+,1];
n1=N_total-n2;
print N_total n1 n2;

/* Creation of all possible permutations */
```

```
start perm(n,n_1);
  matrix = shape(0,(gamma(n+1)/
  (gamma(n_1+1)*gamma(n-n_1+1))),n);
  index = 1;
  vektor=shape(-1,1,n);
  pos = 1;
  ok = 1;
  do while(ok=1);
   if pos > n then do;

   if vektor[,+] = n_1 then do;
    matrix[index,]= vektor;
    index = index + 1;
   end;

  pos = pos-1;
 end;
else do;
  if vektor[,pos] < 1 then do;
   vektor[,pos] = vektor[,pos]+1;
   pos = pos+1;
  end;

  else do;
   vektor[,pos]=-1;
   pos = pos-1;
  end;
end;

if pos < 1 then ok = 0;
end;

return (matrix);
finish;

permutations = perm(N_total,n1);
P=Nrow(permutations);

/* Calculation of the BWS test statistic */
start test_sta(R1, R2, N_total, n1, n2);

b=R1;
R1[,rank(R1)]=b;
b=R2;
R2[,rank(R2)]=b;
```

```
i=1:n1;
j=1:n2;
Bx=(1/n1)#sum( (R1-(N_total/n1)#i)##2/
   ( (i/(n1+1))#(1-(i/(n1+1)))#((n2#N_total)/n1) ) );
By=(1/n2)#sum( (R2-(N_total/n2)#j)##2/
   ( (j/(n2+1))#(1-(j/(n2+1)))#((n1#N_total)/n2) ) );
B=0.5 # (Bx+By);

return (B);
finish;

/* Carrying out the tests */

Tab=REPEAT(T(ranks),P,1);

R1=choose(permutations=0,.,Tab);
R2=choose(permutations=1,.,Tab);

R1g=R1[loc(R1^=.)];
R2g=R2[loc(R2^=.)];

R1z=shape(R1g,P, n1);
R2z=shape(R2g,P, n2);

test_st0=
   test_sta(T(ranks[1:n1]),T(ranks[(n1+1):N_total]),
   N_total, n1, n2);
Pval=0;

do i=1 to P by 1;
B = test_sta(R1z[ i , ], R2z[ i , ], N_total, n1, n2);
if B >= test_st0 then Pval=Pval+1;
end;

Pval=Pval/P;

/* Definition of output */
x=(Pval || test_st0 || P);
cols={P_value test_statistic total_Perms};
print x[colname=cols];

/* optional: Creation of an output dataset called results */
CREATE results FROM x[colname=cols];
APPEND FROM x;
```

```
CLOSE results;
/*********************************************************/

quit;
%MEND Permtest;
```

In order to apply this macro, the SAS dataset needs the variables group and count (in this order). The dichotomous variable group must have the two values 0 and 1 as codes for the two groups. Thus, the following data step can be used:

```
DATA example1;
  INPUT group count @@;
CARDS;
0 90 0 110 0 118 0 121
1 12 1 22 1 34 1 95
;
```

Then, the macro can be invoked with the following statement, %Permtest(example1); . Please note that further programs in SAS/IML exist that can carry out permutation tests (Gefeller and Bregenzer, 1994; Berry, 1995b). Streitberg and Röhmel (1987) suggested a shift algorithm to perform an exact permutation test in order to compare two independent samples.

When an approximate permutation test is desired, that is, only a random sample out of all possile permutations should be considered, one can use, for example, the following SAS/IML program of Good (2001, p. 207). The program listed here generates a random permutation, Y denotes the observed data, and Ystar the generated permutation. Note that a typing error of Good (2001, p. 207) is corrected here: ranuni instead of randuni.

```
proc iml;
  Y={11, 13, 10, 15, 12, 45, 67, 89};
  n=nrow(Y);
  U=ranuni(J(n,1, 3571));    *3571 is the seed value;
  I=rank(U);
  Ystar=Y(|I,|);
  print Ystar;
quit;
```

An alternative way for an approximate permutation test is to use the shuffle algorithm of Chen and Dunlap (1993, p. 409); this program does not need SAS/IML:

```
%let nop=9999; *Number of permutations;

DATA shuffle (KEEP=sample group count);
  ARRAY temp{*} S1-S800;
```

```
*if the total sample size N > 800, replace 800 by N;
DO sample=1 TO &nop;
DO i=1 to obn;
temp(i)=i;
END;
DO j=1 TO obn;
k=int(ranuni(0)*(obn-j+1))+j;
index=temp(k);
temp(k)=temp(j);
temp(j)=index;
set example1(keep=count) point=index;
set example1(keep=group) point=j nobs=obn;
OUTPUT;
END; END; STOP;
RUN;
```

2.4 Comparison of the Three Tests

We now compare the three tests discussed above regarding size and power. We consider the permutation tests. If more than 100,000 permutations were possible—as for example for $n_1 = n_2 = 10$—the FPP test is performed based on 100,000 randomly selected permutations.

The actual type I error rates of the tests are displayed in Table 2.7. This actual level, also called size of the test, can be determined exactly for the rank tests. The reason is that the values needed to carry out the tests, that is, the ranks 1 to N, do not depend on the observed data. Note that continuous distributions are assumed so far, thus there cannot be ties within the data. However, the permutation null distribution of the FPP test statistic does depend on the observed values. Therefore, the actual level is simulated for this test. We present results for the following distributions: uniform distribution on (0,1), standard normal distribution, χ^2 distribution with df $= 3$, and exponential distribution with rate parameter $\lambda = 1$.

A permutation test always guarantees that the actual level is not larger than the nominal one. Some values in Table 2.7, however, are slightly larger than 0.05. This is caused by the simulation error; the simulated estimation varies around the true actual level, which is not above 0.05. The size of the permutation tests FPP and BWS is very close to the nominal significance level; these tests are hardly conservative. In contrast, the WMW permutation test is more conservative. Analogous results occur for significance levels other than 0.05. The WMW test is conservative because its test statistic is discrete. The BWS statistic B is "less discrete"; that is, it has more mass points. For example, for $n_1 = n_2 = 5$, there are $\binom{10}{5} = 252$ different permutations. The

TABLE 2.7: The size of the three permutation tests WMW, BWS, and FPP for different sample sizes ($\alpha = 0.05$)

n_1	n_2	WMW Test	BWS Test	—— Fisher-Pitman Permutation Test ——			
				Uniform	Normal	χ^2 (3 df)	Expon.
5	5	0.0317	0.0476	0.049	0.048	0.048	0.051
6	6	0.0411	0.0498	0.049	0.046	0.045	0.050
7	7	0.0379	0.0490	0.051	0.052	0.050	0.051
8	8	0.0499	0.0499	0.052	0.051	0.051	0.051
9	9	0.0400	0.0499	0.047	0.051	0.050	0.050
10	10	0.0433	0.0500	0.051	0.048	0.050	0.050
8	5	0.0451	0.0497	0.054	0.050	0.043	0.053
9	7	0.0418	0.0500	0.051	0.048	0.050	0.051
10	5	0.0400	0.0500	0.049	0.052	0.049	0.050

rank sum W, however, can take twenty-six different values only, in contrast to the statistic B with forty-two different mass points. This difference is even bigger for larger sample sizes: There are 184,756 permutations in the case of $n_1 = n_2 = 10$ with 11,833 different values of B, but only 101 mass points for W.

It should be noted that the critical values are determined in the usual way, that is, so that the significance level α is not exceeded. Randomized tests that need an additional auxiliary Bernoulli trial are not considered in this book as they are not acceptable for statistical practice (Mehta and Hilton, 1993; Senn, 2007). The result that the WMW permutation test is hardly conservative for $n_1 = n_2 = 8$ and $\alpha = 0.05$ is caused by the pattern of the steps of the distribution function. For $n_1 = n_2 = 8$, this distribution function of the rank sum is extremely close to 0.975 between two steps.

As the tests guarantee the significance level, a test can be selected solely based on power (Kennedy, 1995). One may assume that the FPP test is more powerful than the rank tests because the complete information of the data is utilized. However, Keller-McNulty and Higgins (1987), van den Brink and van den Brink (1989), as well as Tanizaki (1997) demonstrated that the WMW test can have a larger power than the FPP test. This is the case when the distribution is asymmetric and has tails much heavier than those of the normal distribution. Moreover, Rasmussen (1986) showed that the WMW test is more powerful than the FPP test in case of contaminated normal distributions.

Thus, rank tests such as the WMW test are preferable in the case of heavy tails. Moreover, the BWS test accentuates the tails due to the weighting mentioned above. Therefore, it is of interest to investigate the power of the different rank tests in comparison to the FPP test (Neuhäuser and Senske, 2004). Table 2.8 shows the power of the tests for the distributions already considered in Table 2.7. The values were shifted by $\theta = f \cdot \tilde{\theta}$ in one of the two

TABLE 2.8: The simulated power of the three permutation tests FPP, WMW, and BWS for different distributions

$\tilde{\theta}^a$	Test	Uniform	Normal	χ^2 (3 df)	Expon.	CN1	CN2
0.5	FPP	0.15	0.19	0.17	0.15	0.13	0.13
	WMW	0.13	0.17	0.18	0.19	0.22	0.20
	BWS	0.13	0.18	0.20	0.21	0.24	0.22
1.0	FPP	0.47	0.56	0.47	0.41	0.38	0.32
	WMW	0.41	0.51	0.52	0.49	0.48	0.51
	BWS	0.42	0.52	0.58	0.56	0.59	0.61
1.5	FPP	0.83	0.89	0.76	0.67	0.73	0.53
	WMW	0.75	0.85	0.80	0.75	0.62	0.69
	BWS	0.76	0.86	0.85	0.82	0.77	0.82

$^a\theta = f \cdot \tilde{\theta}$ with $f = 4/15$ (uniform distribution), $f = 0.7$ (exponential distribution), $f = 1$ (normal distribution), and $f = 2$ (χ^2-, CN1 and CN2 distributions).

groups, where f was empirically chosen to obtain powers of comparable size for the different distributions. Because of the results presented by Rasmussen (1986), two contaminated normal distributions are also included. The values are simulated as follows: A value comes from a standard normal distribution with a probability of 0.7, and with a probability of 0.3 a value comes from a normal distribution with mean 5 and standard deviation 0.5 (CN1) and 4 (CN2), respectively. Note that contaminated normal distributions are not only of theoretical interest, but are also often useful models for real data (Bradley, 1977).

As known from published studies, the WMW test is more powerful than the FPP test for asymmetric distributions as well as in the case of heavy tails. The power of the BWS test is in these cases even higher than that of the WMW test; the difference in power is sometimes large. When the FPP test has the largest power (uniform and normal distributions), the difference in power between the tests BWS and WMW is small. Hence, it can be suggested to apply the BWS test rather than the WMW test. Even when the underlying distribution is unknown, the BWS test is a good choice. The reason is that the difference in power is only small in the cases where the FPP test is more powerful. In contrast, the BWS test can be much more powerful than the FPP test, as can be observed, for example, for the CN2 distribution (Table 2.8) and the Cauchy distribution (Neuhäuser and Senske, 2004). Analogous results occur for unbalanced sample sizes, that is, when $n_1 \neq n_2$.

Because of the distinct differences in power in the case of heavy tails and the relatively small differences in the case of light tails, the BWS test can also be recommended for situations where the underlying distribution is unknown but can be assumed to be symmetric. Note that heavy tails are common in practice; they could be observed in 49% of the data sets investigated by Micceri (1989). Heavy tails are also common, for example, for emission data (Freidlin et al., 2003) and in finance (Hall and Yao, 2003).

An assumption of symmetry can be justified, for example, when the difference between two random variables is considered. The difference between two exchangeable random variables has a symmetric distribution. Thus, when group membership is assigned independently and at random, the paired differences have a symmetric distribution (Freidlin et al., 2003). In general, the random variables $Z_1, \ldots, Z_n$ are exchangeable if any permutation has the same joint distribution function (Brunner and Munzel, 2002, p. 199).

It was mentioned previously that the difference in power between the WMW and the BWS tests is small for uniform and normal distributions. However, when the sample sizes $n_1 = n_2 = 10$ are considered, as in Table 2.8, the WMW test is conservative with a size of 0.0433. Therefore, it would be useful to compare the power for $n_1 = n_2 = 8$. For these sample sizes there is no difference in the actual level; both tests are hardly conservative, with a size of 0.0499 (see Table 2.7). The difference in power is still small (Neuhäuser, 2005a), but now, in contrast to the results of Table 2.8, the WMW can be slightly more powerful than the BWS test. This can be observed for normally distributed data. When the location shift is $\theta = 1.5$, the estimated power of the WMW test is 0.775, whereas it is 0.759 for the BWS test ($n_1 = n_2 = 8$, $\alpha = 0.05$).

For $n_1 = n_2 = 8$, the power comparison is fair because there is no difference in size. An alternative way is to consider $n_1 = n_2 = 10$ as in Table 2.8, but to set the nominal level to 0.0433. That is the actual size of the WMW test. Because the BWS test can entirely utilize this nominal level as well, the two exact tests WMW and BWS then have identical size. Table 2.9 shows power results for the usual BWS and WMW tests (from Table 2.8) and in addition the power of the BWS test with the nominal, and actual, level 0.0433 ($BWS_{0.0433}$). As expected, the power of the $BWS_{0.0433}$ test is lower than that of the BWS test with $\alpha = 0.05$, but even the $BWS_{0.0433}$ test is often distinctly more powerful than the WMW test. Only for the uniform and the normal distributions is the WMW test slightly more powerful than the $BWS_{0.0433}$ test. Thus, the BWS test has not only the advantage of being less conservative in comparison with the WMW test, but it is also more powerful for many distributions. And the difference in power is not (only) caused by the difference in size (Neuhäuser, 2005a).

The distribution of W is less discrete for large sample sizes; therefore the WMW test is less conservative. Because the difference in power between the tests BWS and WMW is not only due to the difference in size, the BWS test has a power benefit for larger sample sizes as well. Table 2.10 in Section 2.6 shows that for $n_1 = n_2 = 50$.

For large sample sizes, the powers of the FPP test and the t test are similar; in fact, the powers are asymptotically equal (Lehmann, 2006, p. 106). According to the results about the asymptotic relative efficiency between the

TABLE 2.9: The simulated power of permutation tests for different distributions, the power of the $BWS_{0.0433}$ test, that is, a BWS test with the nominal level 0.0433, is displayed in addition to results already presented in Table 2.8

Uniform $(\theta = \frac{6}{15})$	Normal $(\theta = 1.5)$	χ^2 (3 df) $(\theta = 3)$	Expon. $(\theta = 1.05)$	CN1 $(\theta = 3)$	CN2 $(\theta = 3)$
WMW test:					
0.75	0.85	0.80	0.75	0.62	0.69
BWS test:					
0.76	0.86	0.85	0.82	0.77	0.82
$BWS_{0.0433}$ test:					
0.74	0.84	0.83	0.80	0.72	0.80

WMW and the t test (see e.g. Lehmann, 2006, pp. 78–81; as well as Blair and Higgins, 1981, for contaminated normal distributions), it could be expected that the WMW test is more powerful than the FPP test for the exponential distribution or some contaminated normal distributions, at least for large samples. The results presented here show that these results also hold in the case of small samples.

2.5 One-Sided Alternatives

In the previous section the tests were compared for the two-sided alternative $H_1: \theta \neq 0$. Nevertheless, the different tests considered so far might also be used for the one-sided alternatives $H_1^>: \theta > 0$ or $H_1^<: \theta < 0$, respectively. Note that the term *one-tailed* can also be used in place of *one-sided*. The test statistic P of the FPP test, however, cannot detect the direction of a possible difference. Therefore, another statistic such as the difference of means $\bar{X} - \bar{Y}$ should be used when carrying out a one-sided FPP test. For a one-sided WMW test, one can use the rank sum W; this test statistic must be standardized when performing an asymptotic test.

A one-sided test should be applied only if a one-sided alternative is appropriate, for instance when a difference in one direction only is possible because of theoretical a priori information or previous studies. An example is presented in Section 7.5. If the choice between one- and two-sided alternatives is not clear-cut, a two-sided test should be performed in case of doubt. Hence, the two-sided test is the standard, and one deviates from this standard in justified cases only: "we recommend using two-sided tests except in very special circumstances" (Whitlock and Schluter, 2009, p. 143, see also Ruxton and Neuhäuser, 2010a).

When a one-sided permutation test is carried out, only the tail at one side of the permutation null distribution is relevant for the p-value calculation. The one-sided alternative specifies which side must be considered. Let us again study the data of Good (2001). When applying the WMW test we have $W = 25$. Under the one-sided alternative $H_1^>$, the values of group 1 are stochastically larger than those of group 2. The p-value of the one-sided test with this alternative is $P_0(W \geq 25) = 2/70$ (cf. Table 2.3). For the one-sided test with the alternative $H_1^<$, the p-value would be $P_0(W \leq 25) = 69/70$ (cf. Table 2.3).

Some statistics packages, including SAS, if asked to carry out a one-sided test, sometimes report both one-tailed tests but sometimes only report the test (from the two alternatives) that provides the smallest p-value for the data at hand (Ruxton and Neuhäuser, 2010a). An example for the latter case is the SAS procedure NPAR1WAY: Although the SAS programs in Sections 2.1 and 2.2 do not specify one of the two possible one-sided tests, there is one one-sided test only in the output. For instance, the output presented in Section 2.1 includes

`One-Sided Pr >= S     0.0286`

This p-value, specified by the SAS procedure NPAR1WAY, is the smallest of the two possible one-tailed p-values. Obviously, if at all, the one-sided alternative must be specified in advance, that is, before there is any descriptive exploration of the data. Therefore, one has to verify whether or not the specified one-sided test was performed. It is possible to deduce which of the one-tailed tests the software actually implemented. Namely, one can check whether the observed difference shows into the same direction as the specified one-sided alternative. If not, the output presents the wrong one-tailed test.

Due to the squares in the numerators of B_X and B_Y, the BWS statistic B cannot be directly used for testing a one-sided alternative. Therefore, the following modification was proposed (Neuhäuser, 2001a):

$$B_X^* = \frac{1}{n_1} \sum_{i=1}^{n_1} \frac{\left(R_i - \frac{N}{n_1} \cdot i\right) \cdot \left|R_i - \frac{N}{n_1} \cdot i\right|}{\frac{i}{n_1+1} \cdot \left(1 - \frac{i}{n_1+1}\right) \cdot \frac{n_2 N}{n_1}} \quad \text{and}$$

$$B_Y^* = \frac{1}{n_2} \sum_{j=1}^{n_2} \frac{\left(H_j - \frac{N}{n_2} \cdot j\right) \cdot \left|H_j - \frac{N}{n_2} \cdot j\right|}{\frac{j}{n_2+1} \cdot \left(1 - \frac{j}{n_2+1}\right) \cdot \frac{n_1 N}{n_2}}.$$

If $\theta < 0$, that is under the alternative $H_1^<$, the values of group 2 are stochastically larger than those of group 1. In this case, the larger the difference between the groups, the larger is B_Y^*; however, B_X^* decreases with an increasing difference between the groups. Hence,

$$B^* = \frac{1}{2} \cdot (B_Y^* - B_X^*)$$

is suitable for testing H_0 versus $H_1^<$. It would be possible to define B_X^* and B_Y^* without the second multiplier in the numerator (i.e., the multiplier that is always nonnegative due to the absolute values). However, the resulting statistic is more discrete than B^*, that is, it can have less mass points (Neuhäuser, 2002a). Consequently, one should not omit the second multipliers.

When comparing the exact permutation test based on B^* with the one-sided permutation tests WMW and FPP, the results are analogous to those for two-sided tests presented above. Therefore, these results for one-sided tests are not presented here.

In summary, the exact BWS tests, based on B or B^* depending on the chosen alternative, can be recommended for testing location differences in the two-sample location-shift model. This recommendation also holds for multiple test problems (Neuhäuser and Bretz, 2001; Neuhäuser, 2002a).

2.6 Adaptive Tests and Maximum Tests

As mentioned above, there is a large variety of different linear rank statistics. Up to now, we only considered the WMW test, which is relatively powerful for symmetric distributions with medium or large tails. However, in a nonparametric model, the distributions F and G are unknown; thus, they might be asymmetric and/or might have short or heavy tails. In these cases, other scores than $g(i) = i$ are useful. An ideal way, at least at first glance, would be to use the data in order to estimate the scores of the locally most powerful rank test. The resulting locally optimal score function is

$$g_{\text{opt}}(i, f) = \frac{-f'\left(F^{-1}\left(\frac{i}{N+1}\right)\right)}{f\left(F^{-1}\left(\frac{i}{N+1}\right)\right)},$$

where F is the distribution function that holds for both groups under H_0, and f denotes the corresponding density function. Unfortunately, this fine-adaptive procedure, explained in detail by Behnen and Neuhaus (1989), requires very large sample sizes (Büning and Trenkler, 1994, p. 308; Brunner and Munzel, 2002, p. 239). Thus, its suitability for statistical practice is very limited.

An alternative method is to choose between a few preselected score functions. The choice is done by a selector statistic that must be, under H_0, independent of all possible test statistics. This approach, introduced by Hogg (1974) and further developed by Büning (1991), is based on the following lemma:

Lemma 1:

(i) Let $\mathcal{F}$ denote the class of distributions under consideration. Suppose that each of k tests $T_1, \ldots, T_k$ with the critical regions $C_1, \ldots, C_k$ is distribution-free over $\mathcal{F}$, that is,

$$P_0(T_i \in C_i \mid F) \le \alpha \text{ for each } F \in \mathcal{F} \text{ and } i = 1, \ldots, k.$$

(ii) Let S be some statistic (called a selector statistic) that is, under H_0, independent of $T_1, \ldots, T_k$ for each $F \in \mathcal{F}$. Suppose we use S to decide which test T_i to conduct. Specifically, let M_S denote the set of all values of S with the following decomposition:

$$M_S = D_1 \cup D_2 \cup \cdots \cup D_k, \ D_i \cap D_j = \emptyset \text{ for } i \neq j,$$

so that $S \in D_i$ corresponds to the decision to use test T_i.

The overall testing procedure is then defined by:
If $S \in D_i$, then reject H_0 if $T_i \in C_i$. This two-staged adaptive test is distribution-free under H_0 over the class $\mathcal{F}$; that is, it maintains the level α for each $F \in \mathcal{F}$. $\qquad \Box$

The proof of this lemma is given, for example, by Randles and Wolfe (1979, p. 388).

Here, we shall consider an adaptive test based on four linear rank statistics. In addition to the Wilcoxon rank sum test, the following three tests will be included (Büning, 1996, 1997):

• Gastwirth test (short tails):

$$g(i) = \begin{cases} i - \frac{N+1}{4} & \text{for} \quad i \le \frac{N+1}{4} \\ 0 & \text{for} \quad \frac{N+1}{4} < i < \frac{3(N+1)}{4} \\ i - \frac{3(N+1)}{4} & \text{for} \quad i \ge \frac{3(N+1)}{4}, \end{cases}$$

• Hogg-Fisher-Randles (HFR) test (right-skewed distributions, Hogg et al., 1975):

$$g(i) = \begin{cases} i - \frac{N+1}{2} & \text{for} \quad i \le \frac{N+1}{2} \\ 0 & \text{for} \quad i > \frac{N+1}{2}, \end{cases}$$

• LT test (heavy tails):

$$g(i) = \begin{cases} -\left(\left[\frac{N}{4}\right] + 1\right) & \text{for} \quad i < \left[\frac{N}{4}\right] + 1 \\ i - \frac{N+1}{2} & \text{for} \quad \left[\frac{N}{4}\right] + 1 \le i \le \left[\frac{3(N+1)}{4}\right] \\ \left[\frac{N}{4}\right] + 1 & \text{for} \quad i > \left[\frac{3(N+1)}{4}\right]. \end{cases}$$

Above, in the parenthesis that type of distribution is indicated for which the test has high power; $[x]$ denotes the floor function, that is, the largest integer which is smaller than or equal to x. The tests were proposed for different degrees of skewness and tailweight. Therefore, measures for these two characteristics, skewness and tailweight, are suitable selector statistics.

Because the order statistics of all N values is, for independent and identically as well as continuously distributed random variables, independent of the rank vector (see, e.g., Büning and Trenkler, 1994, p. 56), it is possible to estimate the measures from the order statistics. Here, we use measures introduced by Hogg (1974); namely,

$$\hat{Q}_1 = \frac{\hat{U}_{0.05} - \hat{M}_{0.5}}{\hat{M}_{0.5} - \hat{L}_{0.05}}$$

as a measure for skewness, and

$$\hat{Q}_2 = \frac{\hat{U}_{0.05} - \hat{L}_{0.05}}{\hat{U}_{0.5} - \hat{L}_{0.5}}$$

as a measure for tailweight; $\hat{L}_\gamma$, $\hat{M}_\gamma$, and $\hat{U}_\gamma$ denote the arithmetic means of the smallest, medium, and largest γN order statistics, respectively, of the pooled sample. If γN is not an integer, corresponding proportions can be used. The following examples from Büning and Trenkler (1994, p. 306) illustrate the use of such proportions. In these examples the total sample size is $N = 50$, and $x_{(1)}, x_{(2)}, \ldots, x_{(N)}$ denote the ordered pooled sample with the observations from both groups:

$$\hat{L}_{0.05} = \frac{x_{(1)} + x_{(2)} + 0.5x_{(3)}}{2.5}$$

$$\hat{L}_{0.5} = \frac{x_{(1)} + \cdots + x_{(25)}}{25}$$

$$\hat{M}_{0.5} = \frac{0.5x_{(13)} + x_{(14)} + \cdots + x_{(37)} + 0.5x_{(38)}}{25}$$

$$\hat{U}_{0.05} = \frac{0.5x_{(48)} + x_{(49)} + x_{(50)}}{2.5}$$

$$\hat{U}_{0.5} = \frac{x_{(26)} + \cdots + x_{(50)}}{25} .$$

The selector is a bivariate statistic: $S = (\hat{Q}_1, \hat{Q}_2)$. For a symmetric distribution Q_1 equals 1, for a distribution skewed to the right [left] we have $Q_1 > 1$ [< 1]. The measure Q_2 increases with the tailweight. According to Büning (1996), an adaptive test, called ADA in the following, can therefore be defined as (see also Figure 2.1):

ADA:

If $\hat{Q}_1 \leq 2$, $\hat{Q}_2 \leq 2$	apply the Gastwirth test,
if $\hat{Q}_1 \leq 2$, $2 < \hat{Q}_2 \leq 3$	apply the WMW test,
if $\hat{Q}_1 > 2$, $\hat{Q}_2 \leq 3$	apply the HFR test, and
if $\hat{Q}_2 > 3$	apply the LT test.

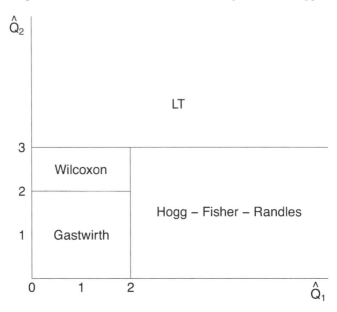

FIGURE 2.1: The adaptive schema for the adaptive test called ADA.

An adaptive test such as ADA does not need sample sizes as large as a fine-adaptive test. However, the sample sizes must not be too small for a reliable selection; that is to avoid too many misclassifications by the selector. According to simulations presented by Hill et al. (1988), the sample size should be at least 20 per group (see also Büning, 1991, p. 238). Therefore, at first we investigate the adaptive test ADA for a sample size ($n_1 = n_2 = 50$) larger than that used in the previous tables.

Table 2.10 displays the size as well as the power of the adaptive test ADA, the underlying single tests, and the BWS test. The single tests, including the BWS test, were performed as approximate permutation tests based on 40 000 permutations in each case. When using the asymptotic distributions, the results are similar. One can see that the adaptive test stabilizes the power; that is, for the different distributions, the power of the adaptive test is always higher than that of the worse test. However, the adaptive test is never the best test.

For small sample sizes, a maximum test is a good alternative to the adaptive test. Such a maximum test uses the maximum of the absolute values of the different standardized statistics as its test statistic (Neuhäuser et al., 2004). Hence, this test does not need any selector that may prove unreliable in the case of small samples. The maximum test called MAX is based on the same four statistics as the adaptive test ADA.

The maximum test MAX is preferable to the adaptive test ADA when sample sizes are small (Neuhäuser et al., 2004). Table 2.11 presents the size

TABLE 2.10: The simulated size and power of the adaptive test ADA, the underlying single test, and the BWS test (approximate permutation tests based on 40,000 permutations) for different distributions

Test	Uniform	Normal	t (2 df)	t (3 df)	χ^2 (3 df)	Expon.
Size ($\tilde{\theta}^a = 0$)						
WMW	0.048	0.051	0.050	0.049	0.053	0.047
Gastw.	0.050	0.050	0.051	0.049	0.054	0.050
HFR	0.046	0.050	0.048	0.046	0.053	0.046
LT	0.047	0.052	0.049	0.050	0.051	0.047
BWS	0.046	0.050	0.048	0.047	0.052	0.047
ADA	0.049	0.051	0.049	0.049	0.052	0.048
Power for $\tilde{\theta} = 0.2$						
WMW	0.14	0.17	0.18	0.22	0.20	0.21
Gastw.	0.24	0.16	0.12	0.16	0.25	0.32
HFR	0.12	0.14	0.15	0.18	0.27	0.32
LT	0.10	0.16	0.19	0.22	0.15	0.14
BWS	0.15	0.16	0.18	0.21	0.22	0.31
ADA	0.22	0.17	0.19	0.22	0.25	0.27
Power for $\tilde{\theta} = 0.4$						
WMW	0.43	0.50	0.56	0.65	0.56	0.57
Gastw.	0.67	0.46	0.33	0.45	0.68	0.76
HFR	0.34	0.41	0.46	0.54	0.72	0.79
LT	0.27	0.45	0.60	0.66	0.44	0.42
BWS	0.51	0.48	0.55	0.63	0.70	0.83
ADA	0.56	0.50	0.60	0.66	0.68	0.67
Power for $\tilde{\theta} = 0.6$						
WMW	0.74	0.83	0.88	0.94	0.87	0.87
Gastw.	0.94	0.79	0.60	0.77	0.93	0.96
HFR	0.64	0.73	0.79	0.87	0.96	0.97
LT	0.52	0.78	0.91	0.95	0.76	0.73
BWS	0.86	0.81	0.88	0.93	0.96	0.99
ADA	0.79	0.82	0.91	0.94	0.92	0.90

[a] $\theta = f \cdot \tilde{\theta}$ with f as in Table 2.8, and $f = 1.5$ for the t distributions with $df = 2$ and 3.

and power for different distributions and $n_1 = n_2 = 10$. For both new tests, ADA and MAX, the size is not larger than the nominal significance level. However, the adaptive test ADA is conservative. Some of the single tests (WMW, Gastwirth and HFR) are conservative too.

A maximum test is also recommended for larger samples. In the case of $n_1 = n_2 = 50$, there is no clear winner when the adaptive test and the maximum test are compared, according to simulations presented by Neuhäuser et al. (2004). An adaptive test is relatively good when sample sizes are large because the selector can work reliably. Nevertheless, a maximum test has the advantages that neither a selector statistic nor an arbitrary decomposition are needed. The latter defines which test should be performed for which values of the selector.

In the mid-1990s, Weerahandi (1995, p. 78) wrote: "Until recently, most of the applications involving nonparametric tests were performed using asymptotic approximations." Therefore, predominant in the literature are adaptive tests that carry out an asymptotic test after the selection. Obviously, permutation tests can also be used after the selection, as done above for the results presented in Tables 2.10 and 2.11: The selector statistic is calculated once and decides which test to perform, and the selected test is carried out based on the permutation null distribution. However, when using a permutation test, another approach is possible: The selector may be calculated for each permutation. This approach is much more flexible, as we discuss below.

According to lemma 1, an adaptive test requires a selector statistic that is, under H_0, independent of all possible test statistics. As mentioned above, the order statistics are independent of the rank vector for a *continuous* distribution and independent and identically distributed random variables. Therefore, rank tests can be used together with selectors based on the order statistics. It is not allowed to compute the selectors $\hat{Q}_1$ and $\hat{Q}_2$ separately for the two groups, and to use the mean values for selection. In this case, the rank statistics are not independent of the selector (Büning, 1991, p. 253). Moreover, as emphasized above, one has to assume continuous distributions (see also Brunner and Munzel, 2002, p. 239).

When the test chosen by the selector is carried out as a permutation test, lemma 1 is needed and therewith the assumption that the underlying distributions are continuous. However, when the selector is calculated for each permutation, lemma 1 is no longer required. The test statistic is defined as

$$T_{APT} = \sum_{i=1}^{k} I(S \in D_i) \cdot T_i \,,$$

where $I(.)$ denotes the indicator function. With this statistic T_{APT}, a permutation test can be performed as usual, and neither the independence between S and the T_i nor the continuity of the underlying distribution is necessary, in contrast to any test based on the lemma. Furthermore, test statistics other than rank statistics, such as P, could be used in the adaptive test.

TABLE 2.11: The simulated size and power of the adaptive test ADA, the maximum test MAX, the underlying single test, and the BWS test (exact permutation tests) for different distributions

Test	Uniform	Normal	χ^2 (3 df)	Expon.	CN1	CN2
Size ($\tilde{\theta}^{a} = 0$)						
WMW			0.043[b]			
Gastwirth			0.042[b]			
HFR			0.043[b]			
LT			0.048[b]			
BWS			0.050[b]			
ADA	0.043	0.045	0.043	0.042	0.040	0.045
MAX			0.050[b]			
Power for $\tilde{\theta} = 1.0$						
WMW	0.41	0.51	0.52	0.49	0.48	0.51
Gastwirth	0.57	0.43	0.42	0.37	0.58	0.27
HFR	0.33	0.41	0.67	0.67	0.66	0.70
LT	0.35	0.49	0.52	0.49	0.44	0.56
BWS	0.42	0.52	0.58	0.56	0.59	0.61
ADA	0.46	0.48	0.56	0.56	0.63	0.63
MAX	0.50	0.50	0.65	0.64	0.69	0.65
Power for $\tilde{\theta} = 1.5$						
WMW	0.75	0.85	0.80	0.75	0.62	0.69
Gastwirth	0.88	0.75	0.60	0.51	0.69	0.38
HFR	0.64	0.75	0.91	0.89	0.83	0.90
LT	0.68	0.84	0.81	0.75	0.58	0.74
BWS	0.76	0.86	0.85	0.82	0.77	0.82
ADA	0.76	0.83	0.80	0.76	0.66	0.78
MAX	0.84	0.84	0.90	0.87	0.88	0.88

[a]$\theta = f \cdot \tilde{\theta}$ with f as in Table 2.8.
[b]Based on the entire permutation null distribution (not simulated).

When using the selectors $\hat{Q}_1$ and $\hat{Q}_2$, the best test is not always selected in practice. However, with the flexibility offered by the new approach with T_{APT}, the independence to the selector is no longer needed, and one can use the standardized test statistics T_i themselves as selectors. To be precise, one can perform a permutation test based on the following statistic:

$$T_{APT2} = \sum_{i=1}^{k} I\left(T_i = \max(T_1, \ldots, T_k)\right) \cdot I\left(T_i > T_j \; \forall \, j < i\right) \cdot T_i \, .$$

The second indicator function is needed because two (or more) statistics could have an equal, and maximal, value $\max(T_1, \ldots, T_k)$. The statistic T_{APT2} is the maximum of the single test statistics. Hence, a maximum test may be regarded as an adaptive permutation test (Neuhäuser and Hothorn, 2006). Therefore, the concept of adaptive tests is more general; maximum tests can be integrated within the theory of adaptive tests. And a maximum test has the advantage that neither a selector statistic nor continuously distributed data are needed. Furthermore, a maximum test is possible for relatively small sample sizes. In contrast, a classical adaptive test needs a sample size of at least twenty per group to avoid too many misclassifications.

For a two-sided alternative, one should use the absolute values of the standardized test statistics for the maximum test. Obviously, the statistics T_{APT} and T_{APT2} can be defined with absolute values.

When applying a maximum test, it is preferable to base it on the permutation null distribution. Freidlin and Korn (2002) demonstrated that the approximation with the asymptotic distribution of the maximum can be distinctly worse, even when all test statistics are normally distributed. A further alternative is a Bonferroni adjustment (see appendix); however, the resulting test is, due to the correlation of the test statistics, less powerful and therefore not recommendable (Neuhäuser et al., 2004).

In case several tests are available for a test problem, combinations other than the maximum are also possible. For instance, one could use the sum of the different statistics rather than the maximum. A better approach than summing up the statistics is the Maximin Efficiency Robust Test (MERT), which exists under weak conditions and amounts to a linear combination of the different test statistics (Gastwirth, 1966, 1970). The comparison between a MERT and the corresponding maximum test depends on the minimal correlation ρ^* between the tests. In various scenarios, the maximum test turned out to be better if $\rho^* \leq 0.5$ (Freidlin et al., 1999; Gastwirth and Freidlin, 2000; Freidlin et al., 2002; Freidlin and Korn, 2002; Zheng et al., 2002). If $\rho^* \geq 0.7$, there was no noteworthy difference in power between the maximum test and the MERT.

Thus, a maximum test is a good choice in the situation considered here. Even when one narrows the set of possible distributions to the family of all t distributions (including the standard normal distribution with df $= \infty$), we have $\rho^* = 0.656$. This minimal correlation is obtained when considering the two most disparate distributions, that is, the extreme pair with Cauchy and normal scores. A maximum test based on the test statistics with these two scores is more powerful than the corresponding MERT, according to the simulations presented by Neuhäuser et al. (2004).

2.7 Ties

Two or more data points with identical numerical values are called ties. As the distribution functions F and G are assumed to be continuous so far, ties almost surely do not occur. In practice, however, ties frequently occur in a variety of settings (see, e.g., Coakley and Heise, 1996). Even when the underlying distribution is continuous, rounding leads to ties. For instance, reaction times may be measured with a time clock graduated in tenths or hundredths of a second. Because more precise measurements are often not required continuous scales are sometimes discretized (Brunner and Munzel, 2002, p. 2). Moreover, it is an advantage of nonparametric rank tests that they can also be applied to discrete numerical and ordered categorical data, but when continuity has to be assumed, this advantage is lost.

Therefore, we no longer assume that F and G are continuous. Apart from that, all other assumptions made at the start of Chapter 2 are still valid. In particular, the two distributions can differ, if at all, by a location shift only.

As mentioned above in Section 2.6, the "classical" adaptive tests assume continuous distributions because of the requirement that selector and test statistics are independent (Brunner and Munzel, 2002, p. 239). Therefore, those adaptive tests are not considered in this section. For a permutation test based on a maximum, however, ties are no problem.

When ties occur, the assignment of ranks is no longer unique. Often, mean ranks are used in practice. To be precise, those observations with identical values are all given an identical ranking; specifically, they are given the arithmetic mean of the ranks that would be assigned to those observations (Büning and Trenkler, 1994, p. 45). Mean ranks are not only often used in applications, but also have a natural, central meaning on theoretical grounds. On the one hand, the sum and therefore the mean of the ranks only depend on n_1 and n_2; they do not depend on the number or pattern of ties (Brunner and Munzel, 2002, p. 37). On the other hand, mean ranks result automatically in the presence of ties when the normalized version of the empirical distribution function is taken as a basis (Brunner and Munzel, 2002, p. 41).

Other approaches to assigning ranks to tied data are possible. One alternative is to randomly break the ties. However, mean ranks give a more efficient test in comparison to randomly broken ties. For the WMW test, this was shown asymptotically by Putter (1955) and Lehmacher (1976). Tilquin et al.

(2003) as well as Neuhäuser and Ruxton (2009a) confirmed these results using simulation studies for small sample sizes.

In the presence of ties, both statistics W and B can be calculated using mean ranks. The variance, however, is smaller than in the case of no ties. Hence, the variance of W cannot be computed as shown in Section 2.2. Instead, one can use the following more general formula for the variance under the null hypothesis (Hollander and Wolfe, 1999, p. 109):

$$
\mathrm{Var}_0(W) = \frac{n_1 n_2}{12} \left(N + 1 - \frac{\sum_{i=1}^{g}(t_i - 1)t_i(t_i + 1)}{N(N-1)} \right),
$$

where g is the number of tied groups and t_i the number of observations in tied group i. A value not tied to other observations is regarded as a "tied group" with size $t_i = 1$.

When the WMW test statistic W is standardized using this formula, the standardized statistic still follows approximately a normal distribution. However, the rule of thumb to use this approximation, for example $\min(n_1, n_2) \geq 7$, no longer holds. In the presence of ties, the goodness of the approximation depends on the number and on the pattern of ties (Brunner and Munzel, 2002, p. 63). Thus, one can hardly recommend an asymptotic test in the case of ties. This also holds for the asymptotic BWS test because it can be extremely anticonservative when using mean ranks in the presence of ties (Neuhäuser, 2002b).

Permutation tests, however, are conditional tests and can also be carried out in the case of ties; they still guarantee the nominal significance level. Permutation tests only require the exchangeability of the observations (Good, 2000, p. 24). Because independent and identically distributed random variables are exchangeable (Brunner and Munzel, 2002, p. 199), the condition of exchangeability is fulfilled in the situation considered here. It should be explicitly pointed out that exact permutation tests are still possible in the presence of ties because some recommendations in the literature are potentially confusing; examples are listed by Neuhäuser and Ruxton (2009a).

In case of one or more ties, a permutation test can be carried out in the same manner as described above for continuous distributions. However, when ties occur, not all of the $\binom{N}{n_1}$ permutations are different. Nevertheless, one can generate all $\binom{N}{n_1}$ permutations, irrespective of whether all are different or not. Thus, the performance of the permutation test does not change. This holds for both ties within groups and ties between groups.

Let us consider the following data as an example to illustrate a permutation test in case of ties: $n_1 = 3$ observations with the values 7, 5, and 5 in group 1, and the following $n_2 = 5$ observations in group 2: 4, 4, 3, 3, 2 (Onghena and May, 1995).

The Fisher-Pitman permutation test is performed first. The three largest

TABLE 2.12: The exact permutation null distribution of P for the data of Onghena and May (1995)

Possible Value of P	Probability ($=$ proportion within the 56 permutations)
0.375	11/56
0.625	7/56
1.375	8/56
1.625	8/56
2.375	7/56
2.625	5/56
3.375	4/56
3.625	4/56
4.375	1/56
4.625	1/56

values are all in group 1; therefore, the sum $\sum X_i$ has its maximum for the actual data. This permutation occurs only once when formally all $\binom{N}{n_1}$ permutations are generated. Thus, for all other permutations, the sum is smaller because at least one value from group 2 is included in the sum. In total there are $\binom{N}{n_1} = \frac{N!}{n_1!n_2!} = 56$ permutations. For the alternative $H_1^>: \theta > 0$, the one-sided p-value $1/56 = 0.0179$ results.

The permutation null distribution of the statistic P is given in Table 2.12. This distribution is not symmetric; P takes the maximum 4.625 for the actual data. Thus, the two-sided p-value is $1/56$ as well. With the t test one gets the following values: $t = 3.54$, df $= 6$, $p_{one-sided} = 0.0061$, $p_{two-sided} = 0.0122$. Due to the symmetry of the t distribution, the two-sided p-value of the t test is always twice as large as the smallest of the two possible one-sided p-values (George and Mudholkar, 1990).

For the WMW and the BWS test, ranks are needed. The smallest value 2 gets the rank 1. The second-smallest value is 3. This value occurs twice; both get the rank $(2 + 3)/2 = 2.5$. The value 4 also occurs twice; the corresponding rank is $(4 + 5)/2 = 4.5$. The values of group 1 get the ranks 8 and twice $(6 + 7)/2 = 6.5$. Hence, the resulting rank sum in group 1 is $W = 21$. The permutation null distribution is displayed in Table 2.13. Under the null hypothesis, the expectation is $E_0(W) = n_1(N + 1)/2 = 13.5$. Thus, 21, the observed value of W, is as far away from the expectation under H_0 as the value 6. For the two-sided p-value of the WMW test, both the permutations with $W \geq 21$ as well as those with $W \leq 6$ must be taken into account. Therefore, there are two permutations, and the resulting two-sided p-value is $P_0(|W - 13.5| \geq 7.5) = 2/56 = 0.0357$.

When applying the BWS test to the data of Onghena and May (1995), we get $B_X = 2.0954$, $B_Y = 5.4334$, and thus $B = 3.7644$. The corresponding

TABLE 2.13: The exact permutation null distribution of W for the data of Onghena and May (1995)

Possible Value of W	Probability (= proportion within the 56 permutations)
6	1/56
8	4/56
9.5	2/56
10	5/56
...	...
17	5/56
17.5	2/56
19	4/56
21	1/56

TABLE 2.14: The exact permutation null distribution of B for the data of Onghena and May (1995)

Possible Value of B	Probability (= proportion within the 56 permutations)
≤ 1.5	44/56
1.5088	4/56
1.6741	2/56
1.9249	4/56
3.7644	1/56
4.2649	1/56

two-sided p-value of the permutation test according to Table 2.14 is $P_0(B \geq 3.7644) = 2/56$.

Let us now look at the size of the different permutation tests. In Table 2.15 the size is, as in Table 2.7, analytically determined for the rank tests and simulated for the FPP test. Ties were created as follows: First, data were simulated according to continuous distribution functions. Then, for the case of one tied group, the values corresponding to the ranks 5 and 6 are replaced by the arithmetic mean of the two values. For the case of two tied groups, the values corresponding to the ranks 10 and 11 are additionally replaced by their mean; and for the case of three tied groups, the values corresponding to the ranks 15 and 16 are also averaged.

Table 2.15 shows that the size of the tests BWS and FPP hardly differs and is close to α, as for continuous distributions. The size of the WMW test is quite variable; it depends on the location of the steps within the discrete

TABLE 2.15: The size of different permutation tests for 1, 2, and 3 tied groups, respectively

Number of Tied Groups	WMW-Test	BWS-Test	— Fisher-Pitman Permutation Test —			
			Uniform	Normal	χ^2 (3 df)	Expon.
1	0.0452	0.0500	0.050	0.048	0.050	0.050
2	0.0499	0.0500	0.051	0.047	0.050	0.050
3	0.0499	0.0499	0.050	0.048	0.050	0.052

TABLE 2.16: The simulated power different of permutation tests for different distributions and for 1, 2, and 3 tied groups, respectively

Number of Tied Groups	Test	Uniform $(\theta = \frac{6}{15})$	Normal $(\theta = 1.5)$	χ^2 (3 df) $(\theta = 3)$	Expon. $(\theta = 1.05)$
1	FPP	0.83	0.89	0.75	0.67
	WMW	0.75	0.86	0.80	0.75
	BWS	0.76	0.85	0.84	0.82
2	FPP	0.83	0.89	0.75	0.67
	WMW	0.77	0.87	0.81	0.77
	BWS	0.76	0.85	0.84	0.82
3	FPP	0.83	0.88	0.75	0.67
	WMW	0.76	0.87	0.81	0.77
	BWS	0.75	0.85	0.84	0.82

distribution function. As a consequence, this test is conservative in case of one tied group, but hardly conservative at all for two or three tied groups.

The power of the tests in the presence of ties is displayed in Table 2.16. In the presented situations, the influence of ties on the power is marginal. Only the power of the WMW test is slightly larger in the case of two or three tied groups, which can be explained by the larger size.

The Tables 2.15 and 2.16 present situations with up to three tied groups only. However, three tied groups are quite substantial for $N = 20$. Even more ties occur in particular in ordinal data as well as in discrete numerical data when there are only a few possible values. A rank test can be applied in the case of ordinal data as illustrated in the next paragraphs. Additional explanations in the context of ordinal data can be found in Chapter 5.

2.7.1   Example

The Old Testament reports in the first chapter of the Book of the Prophet Daniel an experiment. After Nebuchadnezzar, the king of Babel, conquered Jerusalem, Daniel and three other Israelites were brought to the king's palace. The king allotted them food and wine from the royal table, as part of entry into the king's service. But Daniel was resolved not to defile himself with the king's food or wine. So he begged the chief chamberlain to spare him this defilement. However, the chief chamberlain feared that Daniel and his companion would look wretched by comparison with the other young men of their age.

Daniel suggested a test and said: "Give us vegetables to eat and water to drink. Then see how we look in comparison with the other young men who eat from the royal table." Instead of the Israelites, servants ate from the royal table. After ten days, David and the three other Israelites "looked healthier and better fed than any of the young men who ate from the royal table." The warden was convinced and "continued to take away the food and wine they were to receive, and gave them vegetables" (Daniel 1, 1–15). Sprent and Smeeton (2001, p. 7) wrote about this test: "Although the biblical analysis is informal it contains the germ of a nonparametric ... test."

In order to analyze and interpret this test, two important details are missing. On the one hand, we do not know how many servants participated in that experiment. For the further analysis we assume that four servants were included and ate from the royal table, because four persons, that is, the Israelites including Daniel, avoided that food. On the other hand, the question arises whether there was a difference before the test was performed. The Israelites were "of royal blood and of the nobility, young men without any defect." Maybe they already looked healthier and better fed than servants before they swapped their diet. In order to assure the comparability of groups a randomization would be ideal, each person should have been assigned randomly to one of the diets (groups).

The four Israelites were better fed at the end of the experiment than the assumed four servants. We only know this order of the variable nutritional status: The servants had a worse status, the Israelites a higher one. Therefore, the servants get the ranks $(1 + 2 + 3 + 4)/4 = 2.5$, and the Israelites the ranks $(5 + 6 + 7 + 8)/4 = 6.5$. The resulting rank sum in group 1 (servants) is 10. The actual data are an extreme permutation; a lower value of the rank sum is not possible. In a two-sided test, the permutation where the rank sum in group 2 (Israelites) is 10, must be considered for the p-value computation too. In total, there are $\binom{8}{4} = 70$ permutations; hence, the two-sided exact p-value of the WMW test is $2/70 = 0.0286$, significant at the 5% level. We revisit this biblical example in Chapter 5.

2.7.2 Pseudo-Precision

As already mentioned, mean ranks give an asymptotically more efficient test in comparison to randomly broken ties. It is worthwhile in this context to note an issue that is relevant when performing a rank test and the underlying variable must be computed. For some software packages, for example SAS and R, the difference between log $(3)-\log(1)$ and $\log(6)-\log(2)$ is a very small number such as 2e-16, but not zero, as it should be. Under most circumstances, for example for calculating means and variances, this difference would not matter; however, it matters when assigning ranks. A similar problem occurs when a variable is measured by automated equipment. Such values can often be stored, with pseudo-precision, that is, with many more digits (decimal places) than justified by the precision of the respective measurements. The extra digits arise from the memory space that particular computer programming languages allocate to each numerical value of a specific type. Their effect is in almost all cases utterly trivial. However, one case where this noise does matter is in ranking observations. The consequence of such pseudo-precision are more or less randomly broken ties (see also Neuhäuser et al., 2007; Neuhäuser and Ruxton, 2009a).

Let us consider a fictitious example with $n_1 = n_2 = 6$ (Neuhäuser and Ruxton, 2009a). In group 1, the values 1 and 3 are observed three times each. In group 2, we have two times the value 2, three times the value 4, and once the value 5. Note that there are ties within the groups only. The ranks in group 1 are 2, 2, 2, 7, 7, and 7; hence, the rank sum is 27. The resulting p-value of the exact WMW permutation test is 0.0498.

Now we assume that the values are recorded with pseudo-precision, so that, as a consequence, all the ties get lost. The rank sum is unchanged, and the significance (at $\alpha = 0.05$) of the WMW permutation test gets lost, the p-value now being 0.0649.

If an asymptotic WMW test based on the normal distribution is performed, the significance gets lost again. The reason is that, despite the constant rank sum, the variance of the rank sum increases by the breaking of ties, from 6.101 to 6.245. Hence, the standardized rank sum decreases.

Figure 2.2 displays simulation results showing that an exact WMW test based on mean ranks is more powerful than a test based on randomly broken ties. For this figure, normally distributed data were simulated and rounded to one decimal place. Thus, of course, the original data are known. The power of the exact WMW test based on those original values hardly differs from that based on randomly broken ties (Neuhäuser and Ruxton, 2009a).

When pseudo-precision is possible, it is important to round calculated values to an appropriate number of decimal places before computing the ranks. Neuhäuser and Ruxton (2009a, p. 298) suggested that the "number of decimal places should be selected by the scientist on the basis of their understanding of the precision of measurements involved and not by the means by which computers store approximations to recorded values. Failure to do this can

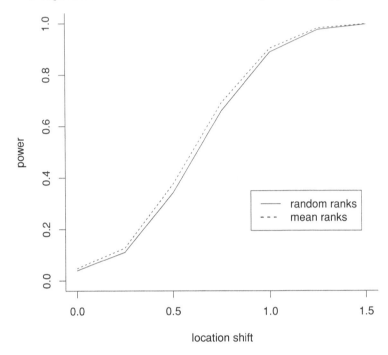

FIGURE 2.2: The simulated size and power of two WMW permutation tests, based on normally distributed data rounded to one decimal place.

lead to artificial reduction in the number of ties in a data set." In this context it should be noted that the SAS documentation also suggests rounding: "[the SAS procedure] PROC NPAR1WAY bases its computations on the internal numeric values of the analysis variables; the procedure does not format or round these values before analysis. When values differ in their internal representation, even slightly, PROC NPAR1WAY does not treat them as tied values. If this is a concern for your data, then round the analysis variables by an appropriate amount before invoking PROC NPAR1WAY" (SAS Institute Inc., 2004, p. 3163).

2.8   Further Tests

Permutation tests can be carried out with many other test statistics. For instance, one can use the difference of medians or trimmed means instead of the difference of means (Efron and Tibshirani, 1993, p. 211). When calculating a trimmed mean, a small prespecified percentage of the largest and smallest values is discarded before calculating the mean. A huge advantage of a permutation test is that it is not required to analytically derive the distribution of the test statistic. The permutation null distribution can be determined for any statistic. For example, one can obtain the permutation null distribution of a maximum without knowing the correlation between the single statistics.

Here, we would like to refer to two further rank statistics. Murakami (2006) introduced a modification of the BWS statistic. Recall that $R_1 < \cdots < R_{n_1}$ (and $H_1 < \cdots < H_{n_2}$, respectively) are the combined-samples ranks of group 1 (group 2) in increasing order. Under the null hypothesis we have

$$\mathrm{E}_0(R_i) = \frac{N+1}{n_1+1} \cdot i$$

and

$$\mathrm{Var}_0(R_i) = \frac{i}{n_1+1}\left(1 - \frac{i}{n_1+1}\right)\frac{n_2(N+1)}{n_1+2},$$

as well as

$$\mathrm{E}_0(H_j) = \frac{N+1}{n_2+1} \cdot j$$

and

$$\mathrm{Var}_0(H_j) = \frac{j}{n_2+1}\left(1 - \frac{j}{n_2+1}\right)\frac{n_1(N+1)}{n_2+2}.$$

These expected values and variances are used in the modified BWS statistic according to Murakami (2006):

$$\tilde{B}^* = \frac{1}{2} \cdot \left(\tilde{B}_X^* + \tilde{B}_Y^*\right), \quad \text{with}$$

$$\tilde{B}_X^* = \frac{1}{n_1}\sum_{i=1}^{n_1}\frac{(R_i - \mathrm{E}_0(R_i))^2}{\mathrm{Var}_0(R_i)} = \frac{1}{n_1}\sum_{i=1}^{n_1}\frac{\left(R_i - \frac{N+1}{n_1+1}\cdot i\right)^2}{\frac{i}{n_1+1}\cdot\left(1 - \frac{i}{n_1+1}\right)\cdot\frac{n_2(N+1)}{n_1+2}} \quad \text{and}$$

$$\tilde{B}_Y^* = \frac{1}{n_2}\sum_{j=1}^{n_2}\frac{(H_j - \mathrm{E}_0(H_j))^2}{\mathrm{Var}_0(H_j)} = \frac{1}{n_2}\sum_{j=1}^{n_2}\frac{\left(H_j - \frac{N+1}{n_2+1}\cdot j\right)^2}{\frac{j}{n_2+1}\cdot\left(1 - \frac{j}{n_2+1}\right)\cdot\frac{n_1(N+1)}{n_2+2}}.$$

When comparing the exact tests based on $\tilde{B}^*$ and B, the modification can be more powerful in the case of unbalanced sample sizes, that is, when $n_1 \neq n_2$. However, there is no clear winner, neither in the case of balanced or unbalanced samples. Obviously, the asymptotic distribution of the modified statistic $\tilde{B}^*$ equals that of the original statistic B (Murakami, 2006).

Zhang (2006) suggested, among others, the following test statistic:

$$Z_C = \frac{1}{N}\sum_{i=1}^{n_1}\ln\left(\frac{n_1}{i-0.5}-1\right)\ln\left(\frac{N}{R_i-0.5}-1\right)$$

$$+ \frac{1}{N} \sum_{j=1}^{n_2} \ln \left(\frac{n_2}{j - 0.5} - 1 \right) \ln \left(\frac{N}{H_j - 0.5} - 1 \right).$$

With this statistic Z_C, one can perform a permutation test. Zhang (2006) did not present an asymptotic distribution. Please note that small values of Z_C give evidence against the null hypothesis.

It should be noted that neither the BWS test and its modification nor the Z_C test were proposed for the location problem. These tests can detect other differences between the groups too, for instance, differences in variability. In this chapter, however, it was assumed that the two groups differ, if at all, by a shift in location. Under this restriction we can use the tests for testing the null hypothesis H_0: $\theta = 0$. In this situation the three tests, the BWS test, its modification, and the Z_C test, can be more powerful than the tests WMW and FPP. A clear winner could not be detected in simulations.

In practice, however, there arises not only the question of which test to perform, but also the question of whether or not the location-shift model is appropriate. When other differences between the groups—as for example different variances—are possible, the assumption $F(t) = G(t - \theta)$ is not justified. The following Chapter 3 focuses on tests that might be appropriate in that case. First, the sample size calculation in the location model is discussed for the example of the WMW test.

2.9   Sample Size Calculation

Before an experiment or a study is performed, the necessary sample size should be calculated. For a statistical test, the probability of a type I error is bounded. This probability is not larger than the significance level α (if the test guarantees the level, that is, if the test is not anticonservative). The probability of a type II error, however, is not bounded, except for the fact that any probability is ≤ 1, and, hence, could be close to 1.

Now, it should be required that the probability of a type II error is not larger than β for a specified $\theta \neq 0$. Hence, two values must be provided: θ and β, where θ is the difference that should be detected with a large probability. This probability is $1 - \beta$, the power (see appendix). According to the international ICH E9 guideline (ICH, 1999), the power must be at least 0.8 for clinical studies. Thus, the probability of a type II error is not larger than 0.2 (for the specified θ).

The aim of a sample size calculation is to determine the sample size required in order to achieve a power of $1 - \beta$ for the specified θ. This principle is illustrated for the one-sided WMW test with the alternative $H_1^>$. For this test, the total sample size N, required to achieve the power $1 - \beta$ at significance level

α, can be approximated with the following formula (Noether, 1987; Hollander and Wolfe, 1999, p. 120):

$$N \approx \frac{(z_{1-\alpha} + z_{1-\beta})^2}{12c(1-c)(\delta - 0.5)^2},$$

where $z_{1-\alpha}$ denotes as above the $(1 - \alpha)$ quantile of the standard normal distribution. In addition, we have $\delta = P(X_i > Y_j)$ and $n_2 = cN$. Thus, $c = 0.5$ results for a balanced design.

As an example, the sample size required for a power of $1 - \beta = 0.9$, with $\alpha = 0.025$, $c = 0.5$, and $\delta = 0.65$, is calculated. Based on the quantiles $z_{0.975} = 1.960$ and $z_{0.9} = 1.282$, we have

$$N \approx \frac{(1.960 + 1.282)^2}{3 \cdot 0.15^2} = 155.7.$$

In order to guarantee the required power of 90%, the result is rounded up. Hence, a total sample size of $N = 156$ resulted. Because of the balanced design, we get $n_1 = n_2 = 78$.

What does the chosen value of $\delta = 0.65$ mean for θ, the shift in location? This is illustrated for the normal distribution: We assume that the observations of both groups are normally distributed with variance 4. Then, the difference $X_i - Y_j$ is normal with mean θ and variance 8. The probability $P(X_i - Y_j > 0)$ equals $\delta = P(X_i > Y_j)$, which is 0.65 if $\theta = 1.09$.

Further, more recent approaches to calculate the power and sample size, respectively, of the WMW test can be found in Shieh et al. (2006) as well as Rosner and Glynn (2009).

2.10 Concluding Remarks

A permutation test is often useful, especially when sample sizes are small and the underlying distributions F and G are unknown. Irrespective of which test statistic is applied, one has to carry out five steps when performing a permutation test. First, a suitable test statistic must be selected, before data collection. This chosen statistic is calculated for the original data (i.e., the actual observations). Then, all possible permutations must be generated, and the test statistic is computed for each permutation. Finally, the p-value is determined as the proportion of permutations that give at least as much evidence in favor of the alternative as the actual data.

As mentioned above, a permutation test can be carried out with a huge variety of possible test statistics. For example, one can use a maximum or the difference between trimmed means. However, a practicing statistician may be confronted with the situation that the desired test is not implemented in standard software. In that case, the possible permutations must be listed; the SAS programs given in Section 2.3 can be used for this task.

In chapter 2 the location-shift model was considered. However, as mentioned above, the assumption $F(t) = G(t - \theta)$ is not always justified. The assumptions needed in Chapter 3 are less restrictive.

Chapter 3

Tests in the Case of Heteroscedasticity

"the assumption of homoscedasticity ... is usually made for simplicity and mathematical ease rather than anything else" (Ogenstad, 1998, p. 497).

In practice, a shift in location is often accompanied by an increase in variability (see, e.g., Singer, 2001). According to Blair and Sawilowsky (1993), this phenomenon is common in toxicological, medical, and epidemiological studies. In a placebo-controlled clinical trial, the placebo group (a placebo is a simulated medical intervention such as an inert tablet) often has a smaller variability due to differences in the response to the therapy (Bender et al., 2007). In genetics, too, variances often increase with increasing means (Jansen, 2001, p. 571). When considering gene expression data, unequal variances between different groups usually occur at least for a part of genes (Demissie et al., 2008). Figure 3.1 shows hypothetical distributions for gene expression for normal and cancerous tissues (cf. Pepe et al., 2003, p. 134). There are differences in both location and variability regarding all three displayed distributions. This was also the case for the empirical distributions observed by Pepe et al. (2003, see their Figure 3). Differences in location are also often accompanied by heteroscedasticity, that is, a difference in variability, in social and behavioral sciences (see, e.g., Neuhäuser, 2001a). Thus, the assumption of a pure shift in location, as made in Chapter 2, might be an oversimplification that is not always appropriate.

The tests discussed in Chapter 2 can be significant with a probability much larger than α when the distributions underlying the two groups differ in variability only. Boik (1987) demonstrated this result for the FPP test, Neuhäuser (2000) for the BWS test, and, for example, Kasuya (2001) for the WMW test.

Under the restriction that the distributions F and G are symmetric, and under the assumption that there is no difference in location, the probability of obtaining a significance in a one-sided asymptotic WMW test has the limit

$$1 - \Phi \left[\frac{z_{1-\alpha}}{\sqrt{12}} \left(\lambda_1 \mathrm{Var}(F(X)) + \lambda_2 \mathrm{Var}(G(Y)) \right)^{-1/2} \right],$$

where Φ is the distribution function and $z_{1-\alpha}$ the $(1 - \alpha)$ quantile of the

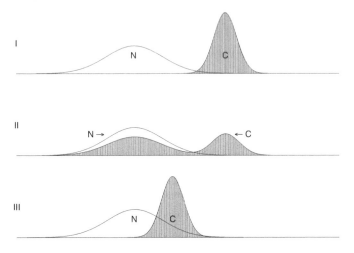

FIGURE 3.1: Hypothetical distributions for gene expression data for normal [N] and cancerous [C] tissues.

standard normal distribution N(0, 1); λ_1 and λ_2 denote the limits for n_1/N and n_2/N, respectively, $0 < \lambda_i < 1$ (Hettmansperger and McKean, 1998, p. 129). Under the additional assumption that $G(x) = F(x/\eta)$, $\eta > 0$, and a balanced design ($n_1 = n_2$), the above limit can also be given as

$$1 - \Phi \left[\frac{z_{1-\alpha}}{\sqrt{12}} \left(\frac{1}{2} \int F^2(\eta t) dF(t) + \frac{1}{2} \int F^2(x/\eta) dF(x) - \frac{1}{4} \right)^{-1/2} \right].$$

This limit for the probability of obtaining a significance in a one-sided asymptotic WMW test is displayed for the standard normal distribution F in Figure 3.2. The limit has its minimum α at $\eta = 1$, irrespective of the distribution F. Hence, the limit is always larger than α for $\eta \neq 1$. The maximum is $1 - \Phi(0.816 z_{1-\alpha})$, for instance 0.09 for $\alpha = 0.05$, and is attained at the boundaries 0 and ∞ (Hettmansperger and McKean, 1998, pp. 129–130). The limit $1 - \Phi(0.816 z_{1-\alpha})$ for the probability of obtaining a significance in a one-sided asymptotic WMW test also holds for a two-sided test, because the rank sum W has a symmetric distribution in case of a balanced design and, hence, a two-sided test can be expressed as a combination of two one-sided tests with significance levels $\alpha/2$.

When the restrictive assumptions, in particular the symmetry of the distributions F and G as well as the equality $n_1 = n_2$, are abandoned, the probability for a significance in the WMW test due to different variabilities can be distinctly larger (Kasuya, 2001). Despite its conservatism, this also holds for the exact WMW test. The probability for a significance is relatively high when the smaller group has the larger variability (see, e.g., Brunner and Munzel, 2000). However, probabilities for a significance, without a difference

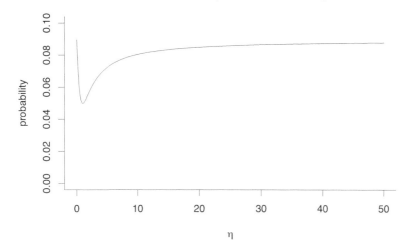

FIGURE 3.2: The limit for the probability of obtaining a significance in a one-sided asymptotic WMW test.

in location, which are much larger than the nominal level α are possible even in case of $n_1 = n_2$, as shown for example by the extensive simulation study of Zimmerman (2003).

For the FPP test, the probability of obtaining a significance solely due to a difference in variability can also be larger than α, as for the WMW test. In the case of a balanced design and symmetric distributions, this probability is not extremely large, as shown for example by the simulations of Hayes (2000). In case of unequal sample sizes, the probability is larger than $\alpha = 5\%$, when the smaller group has the larger variability, and the probability is smaller than 5%, as long as the variability is smaller in the smaller group (Table 3.1; Hayes, 2000). These results are, at least for large sample sizes, not surprising as analogous findings are known for the parametric t test (Boneau, 1960).

In contrast to the FPP and WMW tests the BWS test can detect differences in variance with a relatively large probability (see Figure 2B in Baumgartner et al., 1998). This also holds for the modification $\tilde{B}^*$ proposed by Murakami (2006) and the Z_C test of Zhang (2006). Therefore, the probability of obtaining a significance due to a difference in variability is much larger (see also Table 3.1). However, when the larger group has the larger variability, the probability for a significance in the BWS test is small.

So far wording such as "the probability of obtaining a significance solely due to a difference in variability" has been used. The question arises whether this probability is power or size. The answer depends on the definition of the null and alternative hypotheses.

The alternative can be very general, $F \neq G$; that is, there is a t with $F(t) \neq G(t)$. In this case the alternative is true when there is a difference in variability but none in location. Thus, the probability of obtaining a significance solely due to a difference in variability is power.

TABLE 3.1: Simulated probability of obtaining a significance with the permutation tests FPP, WMW, and BWS, based on normally distributed data with mean 0 in both groups

n_1	n_2	Test	Standard Deviations in the Two Groups			
			1, 1	1, 2	1, 3	1, 4
5	10	FPP	0.05	0.02	0.02	0.01
		WMW	0.04[a]	0.02	0.02	0.03
		BWS	0.05[a]	0.04	0.05	0.06
10	10	FPP	0.05	0.06	0.06	0.07
		WMW	0.04[a]	0.05	0.06	0.07
		BWS	0.05[a]	0.09	0.17	0.26
10	5	FPP	0.05	0.13	0.17	0.20
		WMW	0.04[a]	0.07	0.09	0.09
		BWS	0.05[a]	0.11	0.16	0.19

[a]Based on the entire permutation null distribution (not simulated).

On the other hand, in the so-called Behrens-Fisher problem it is tested whether the locations of the two groups are equal, irrespective of a possible difference in variability. In this case the null hypothesis is true when the only difference between F and G is heteroscedasticity. As shown above, the tests discussed up to now do not control the nominal significance level in the (nonparametric) Behrens-Fisher problem.

Which of the two possibilities for defining the hypotheses is suitable? The answer depends on the problem and the context. When comparing two samples from different populations, as for example in a comparison of testosterone levels between smokers and nonsmokers, the Behrens-Fisher problem can be appropriate (Neuhäuser, 2002c).

When, however, homogenous experimental units are randomized to different groups or treatments, the equality of variances is a characteristic of the null hypothesis "no treatment effect" (Brownie et al., 1990). Unequal variabilities in the two groups can indicate a difference between treatments. In the context of such an informative heterogeneity of variances (Hothorn and Hauschke, 1998, p. 90), it is appropriate to regard differences in location and in variability as treatment effects. This approach leads to a so-called location-scale test. An adjustment for heteroscedasticity, as proposed for the parametric Behrens-Fisher problem for example by Welch (1937), is not appropriate in this case: "Welch's ... method is appropriate when the unequal variances would be expected in the absence of a treatment effect ... [and] particularly inappropriate if the heterogeneity is attributable to a heterogeneous treatment effect" (O'Brien, 1988, p. 60).

Ludbrook and Dudley (1998), already cited above in Section 2.2, showed that randomization of homogenous experimental units is common, at least in biomedical research. The situation is similar in psychology (Hunter and May, 1993). Therefore, location-scale tests might be preferable and will be discussed in the following Section 3.1. Tests for the Behrens-Fisher problem are addressed in Section 3.2.

3.1 Location-Scale Tests

In contrast to Chapter 2, the two distributions F and G might differ not only by a shift in location. Now, the following assumption will be made: $F(t) = G(\theta_1 t - \theta_2)$ for all t, $\theta_1 > 0$, $-\infty < \theta_2 < \infty$. This assumption is clearly less restrictive as differences in variability are possible. Therefore, this assumption might be more often relevant for statistical practice than the pure location-shift model.

The null hypothesis to be tested is now

$$\mathrm{H}_0^{\mathrm{LS}} : \ \theta_1 = 1 \text{ and } \theta_2 = 0,$$

that is, there is no difference between the distributions F and G. If the variances σ_F^2 and σ_G^2 of the distributions defined by F and G, respectively, exist, we have $\theta_1 = \sigma_G/\sigma_F$. Thus, in this case, $\theta_1 = 1$ is equivalent to equality of variances.

Under the alternative $\mathrm{H}_1^{\mathrm{LS}}$ we have $\theta_1 \neq 1$ or $\theta_2 \neq 0$, that is, there is a difference with regard to the scale parameter θ_1 and/or the location parameter θ_2. As in the location-shift model, the null hypothesis in the location-shift model can be expressed as $F = G$. However, the null hypothesis is no longer equal to the hypothesis $p = 1/2$, where p denotes the relative effect (Brunner and Munzel, 2002, p. 17),

$$p = P(X_i < Y_j) + 0.5P(X_i = Y_j).$$

The hypothesis $F = G$ is only a subset of the hypothesis $p = 1/2$ (Brunner and Munzel, 2002, p. 234).

Figure 3.3 displays the power of the three tests FPP, WMW, and BWS, discussed in detail in Chapter 2, under heteroscedasticity, that is, when there are differences in variability. The power decreases for all three tests with increasing variability. The loss in power, however, is less pronounced for the BWS test than for the other two tests. Note that the decline in power of the WMW test is similar when this test is not conservative for $\alpha = 0.05$ and $n_1 = n_2 = 8$ (Neuhäuser, 2005a).

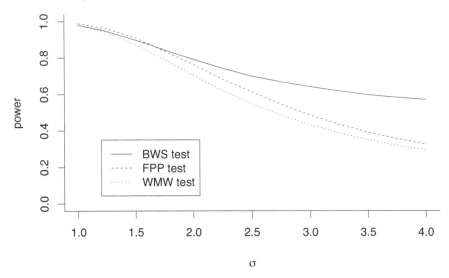

FIGURE 3.3: The simulated power of the three permutation tests FPP, WMW, and BWS, depending on the difference in variance, based on normally distributed data.

Except for the loss in power for increasing variances, all three tests are not suitable as location-scale tests. The reason is that they could not, or at best with a very low power, detect the alternative when sample sizes are unbalanced and the larger group has the larger variability (cf. Table 3.1). This is also the case for the modification $\tilde{B}^*$ of Murakami (2006) as well as the Z_C test of Zhang (2006). For the example of normally distributed data with no difference in means and the standard deviations $\sigma_1 = 1$ and $\sigma_2 = 4$, $n_1 = 5$, $n_2 = 10$, and $\alpha = 0.05$, both tests were significant in less than 10% out of 10,000 simulated data sets. In some similar scenarios the estimated power of the Z_C test was smaller than the nominal significance level. However, it should be noted that the Z_C test is as good as the BWS test in detecting a pure difference in variability in the case of balanced sample sizes.

For the test problem H_0^{LS} versus H_1^{LS}, a stronger sensitivity for scale differences is necessary. Regarding the FPP test, it is obvious that it is primarily sensitive to a location shift, as the test statistic can be expressed as the difference $\bar{X} - \bar{Y}$ (see Section 2.1). For the WMW test, this holds in analogy as the rank sum W arranges the permutations in the same order as the difference between the means of ranks of the two groups.

3.1.1   The Lepage Test

The test introduced by Lepage (1971, see also Hollander and Wolfe, 1999, pp. 169-176) is the commonly used nonparametric location-scale test in practice (Büning, 2002). The test statistic is the sum of the standardized and squared test statistics of the Wilcoxon and the Ansari-Bradley tests. The latter test was introduced as a rank-sum test for dispersion (Ansari and Bradley, 1960).

The test statistic of the Ansari-Bradley test, defined by

$$AB = \frac{1}{2} n_1 (N+1) - \sum_{i=1}^{N} \left| i - \frac{1}{2}(N+1) \right| \cdot V_i ,$$

is asymptotically normal (Hollander and Wolfe, 1999, pp. 144–145). As in Chapter 2, $V_i = 1$ if the i-th smallest of the N observations comes from group 1, else $V_i = 0$.

An exact test can be performed based on the permutation null distribution of AB. This test can detect differences in variability when the larger group has the larger variability. For instance, the (simulated) power is 0.45 when the two samples are normally distributed as $N(0, 1)$ and $N(0, 4^2)$, and for $2n_1 = n_2 = 10$ and $\alpha = 0.05$. The Ansari-Bradley test can be expressed as a linear rank statistic using the scores

$$g_{AB}(i) = \frac{N+1}{2} - \left| i - \frac{N+1}{2} \right| .$$

The test statistic of the location-scale test according to Lepage (1971) is

$$L = \frac{(W - \mathrm{E}_0(W))^2}{\mathrm{Var}_0(W)} + \frac{(AB - \mathrm{E}_0(AB))^2}{\mathrm{Var}_0(AB)} ,$$

where $\mathrm{E}_0(.)$ and $\mathrm{Var}_0(.)$ are the expected values and variances of the test statistics W and AB, respectively, under the null hypothesis. For W, these values are given in Sections 2.2 and 2.7. For the Ansari-Bradley statistic AB, we have, under the assumption of continuous distributions F and G (Lepage, 1971)

$$\mathrm{E}_0(AB) = \begin{cases} \frac{1}{4} n_1(N+2) & \text{if } N \text{ is even} \\ \frac{1}{4} n_1(N+1)^2/N & \text{if } N \text{ is odd} \end{cases}$$

and

$$\mathrm{Var}_0(AB) = \begin{cases} n_1 n_2(N^2 - 4)/(48(N-1)) & \text{if } N \text{ is even} \\ n_1 n_2(N+1)(N^2 + 3)/(48N^2) & \text{if } N \text{ is odd} . \end{cases}$$

In the presence of ties, the expected values are unchanged, but not the variances for which the following formulas can be applied (Hollander and

TABLE 3.2: Colony sizes (defined as the maximum number of workers) of the bumblebee *Bombus terrestris*

Group	Sample Size	Raw Data
High diversity	7	14, 24, 26, 98, 12, 105, 85
Low diversity	12	40, 14, 18, 28, 11, 39,
		17, 37, 52, 30, 65, 35

Source Baer and Schmid-Hempel 1999.

Wolfe, 1999, p. 146):

$$
\mathrm{Var}_0(AB) = \begin{cases} \dfrac{n_1 n_2 (16 \sum\limits_{j=1}^{g} t_j r_j^2 - N(N+2)^2)}{16N(N-1)} & \text{if} \quad N \text{ is even} \\[3em] \dfrac{n_1 n_2 (16N \sum\limits_{j=1}^{g} t_j r_j^2 - (N+1)^4)}{16N^2(N-1)} & \text{if} \quad N \text{ is odd,} \end{cases}
$$

where g is the number of tied groups and t_j the number of observations in tied group j; r_j denotes the mean score g_{AB} for the values of the tied group j. An untied observation is regarded as a "tied group" with $t_j = 1$.

It is evident that a permutation test can be carried out with the statistic L. The alternative for large samples is to use the asymptotic distribution of L. To be precise, L is, under $\mathrm{H}_0^{\mathrm{LS}}$, asymptotically χ^2 distributed with two degrees of freedom, because W and AB are uncorrelated under $\mathrm{H}_0^{\mathrm{LS}}$ (Lepage, 1971; see also Randles and Wolfe, 1979, p. 259).

3.1.2 Example

The Lepage test will be applied to a data set of Baer and Schmid-Hempel (1999) about the colony sizes of the bumblebee *Bombus terrestris*. Young queens were artificially inseminated with sperm of either high (group 1) or low (group 2) genetic diversity. Raw data are displayed in Table 3.2. There are differences in both the means ($\bar{x} = 52.0$ versus $\bar{y} = 32.2$) and the empirical standard deviations (41.9 versus 16.1). There is one tied group; the colony size 14 was observed twice. The ranks of the observations in group 1 are 3.5, 7, 8, 18, 2, 19, and 17. Hence, we get $W = 74.5$, $AB = 26.5$, and the Lepage test statistic

$$
L = \left(\frac{74.5 - 70}{11.827} \right)^2 + \left(\frac{26.5 - 36.842}{5.930} \right)^2 = 0.381^2 + (-1.744)^2 = 3.186 \,.
$$

In order to perform the exact permutation test $\binom{N}{n_1} = \binom{19}{7} = 50,388$, permutations must be generated. The larger the value of the test statistic L, the larger the evidence against the null hypothesis H_0^{LS}. In total, L is 3.186 or larger for 10,506 out of 50,388 possible permutations. Hence, the p-value of the exact Lepage test is $P_0(L \geq 3.186) = 10,506/50,388 = 0.2085$. In case the asymptotic test is carried out despite the small sample sizes, the p-value based on the χ^2 distribution with two degrees of freedom is 0.2033.

3.1.3 Implementation in SAS

The two single test statistics needed for the Lepage test can be computed in SAS using the procedure NPAR1WAY, as presented in Section 2.2 for the WMW test. The Ansari-Bradley test can be applied with the option AB in the PROC NPAR1WAY statement (group and count are the names of the variables, as above in Chapter 2):

```
PROC NPAR1WAY AB;
  CLASS group;
  VAR count;
  EXACT;
  OUTPUT OUT=test_sta;
RUN;
```

In order to compute Lepage's test statistic L, the results of the two single tests should be saved in SAS data sets. With the statement OUTPUT OUT=, a data set called test_sta is created. This data set includes, among others, the test statistic AB and its standardized value, the respective variable names are _AB_ and Z_AB. With the example data of Table 3.2 we have _AB_ = 26.5 and Z_AB = −1.744.

The statement OUTPUT OUT= is also possible when performing the WMW test. Then, the output data set includes, among others, the rank sum W (variable name: _WIL_). In order to receive the standardized value of the rank sum without a continuity correction the square root of the variable _KW_ can be used. In the example, the values are _WIL_ = 74.5 and $\sqrt{\text{_KW_}}$ = 0.381. Contrary to the WMW test, the SAS procedure NPAR1WAY does not apply a continuity correction by default when applying the Ansari-Bradley test.

The test statistic L is Z_AB^2 + _KW_. The p-value of the asymptotic test can be computed as 1 - probchi(L,2). In order to carry out a permutation test, the algorithms presented in Section 2.3 can be used. The exact permutation test with Lepage's statistic—as well as some other nonparametric tests—can also be performed with a SAS/IML program given by Berry (1995b).

The analysis of a data set should not be finished with a significant location-scale test. Neuhäuser and Hothorn (2000) proposed to apply a closed testing procedure (see appendix) after a significant location-scale test in order to separately test for differences in location and variability. In this case, all three tests can be performed with the full significance level α. However, the question arises as to which tests are suitable for step 2 of the closed testing procedure.

Possible tests will be discussed in Sections 3.2 to 3.4. First, some modifications of Lepage's test and the test introduced by Cucconi (1968) will be described.

3.1.4   Weighted and Modified Location-Scale Tests

In Lepage's statistic L, the rank sum W has the same weight as the Ansari-Bradley statistic AB. This weighting is appropriate when no a priori information is available about how big the differences in location and variability are. If it is known that, for example, the location shift is big but the difference in variability is small, it would make sense to give much more weight to W than to AB. However, such a priori information is rare in practice.

A possibility of obtaining some information about the magnitude of the differences is an adaptive interim analysis. Such an interim analysis is possible when a study is subdivided into two (or more) phases. In a two-stage design there is an interim analysis after the first phase. The information obtained within the interim analysis can be used for planning the following phase(s). Neuhäuser (2001b) proposed to weight the location-scale test for the analysis of phase 2 using the results of phase 1. Then, an unweighted test is necessary for the analysis of phase 1 only.

When applying such an adaptive design, the results of k different phases can be combined using Fisher's combination test. Under the particular null hypotheses, the p-values of continuously distributed test statistics are uniformly distributed on the interval $[0, 1]$. If the p-values $p_1, p_2, \ldots, p_k$ are independent, it follows that $-2\ln(p_1 \cdot p_2 \cdots p_k)$ is χ^2 distributed with $2k$ degrees of freedom (see, e.g., Hartung et al., 2008, p. 29). Fisher's combination test is based on this distribution (see also Section 11.2), and can be used to analyze an adaptive design (Bauer and Köhne, 1994). Fisher' combination test can also be applied when the test statistics are discrete, the level of the combination test is maintained in this case (Bauer and Köhne, 1994; see also Section 11.3). Different applications of such an adaptive design can be found, for example, in Neuhäuser (2001c) and Bauer et al. (2001). Note that other combination tests can also be used for the analysis of adaptive designs (see, e.g., Lehmacher and Wassmer, 1999). Further details on combination tests are presented in Sections 11.2 and 11.3.

A weighting of the location-scale test in phase 2 based on an adaptive interim analysis can distinctly increase the power, for example when there is a large difference in variability but only a small shift in location. When there are large differences in both location and variability, a weighting cannot further increase the power (for details, see Neuhäuser, 2001b). In the following we focus on unweighted tests, which are usually needed in practice for one-stage designs and also for phase 1 in adaptive designs.

It was shown in Figure 3.3 that the loss of power is less pronounced for the BWS test, in comparison to the WMW test. Therefore, one might use the BWS statistic B rather than the rank sum W in a modified Lepage test

TABLE 3.3: The simulated power of different permutation tests, based on normally distributed data

Test	Standard Deviations in the Two Groups		
	1, 1	1, 2	1, 3
FPP	0.89	0.53	0.32
WMW	0.85	0.48	0.28
BWS	0.86	0.58	0.48
L	0.76	0.56	0.64
L_M	0.81	0.60	0.64

(Neuhäuser, 2000). Thus, the modified test uses the statistic

$$L_M = \frac{(B - \mathrm{E}_0(B))^2}{\mathrm{Var}_0(B)} + \frac{(AB - \mathrm{E}_0(AB))^2}{\mathrm{Var}_0(AB)} \, .$$

In contrast to L, L_M has no asymptotic χ^2 distribution. Hence, a permutation test is the method of choice. Note that it is not uncommon to apply a test statistic that does not follow any asymptotic standard distribution (North et al., 2002). When generating all permutations, the expected value and the variance of B can be determined based on the permutation null distribution.

There is hardly any difference in size between the Lepage test and its modification. For example, the difference in the actual level is 0.00018 for $n_1 = n_2 = 10$ and $\alpha = 0.05$. However, there can be a gain in power by applying the modification as displayed in Table 3.3 for normally distributed data. For comparison, the power of the single tests FPP, WMW, and BWS is also given. The location-scale tests are less powerful than these single tests when there is a pure difference in location. However, when there is an additional difference in variability, the modified Lepage test is the most powerful of the investigated tests.

When the power mainly comes from a difference in variability (e.g., normal distributions with means 0 and 1.5, and variances 1 and 9) the original Lepage test cannot be improved by the modification with B. This is also the case when there is a difference in variability only, without any shift in location. However, the latter scenario seems to be irrelevant in the applications. Sawilowsky and Blair (1992, p. 358) "never encountered a treatment or other naturally occurring condition that produces heterogeneous variances while leaving population means exactly equal."

The modification can also be more powerful than the original Lepage test in case of nonnormal distributions (Neuhäuser, 2000). Within the simulations for the comparison of different location-scale tests, the exact (Kolmogorov-) Smirnov test was also investigated. This is a test for the general alternative that is the topic of Chapter 4. Here, we only note that the (Kolmogorov-) Smirnov test was less powerful. We do not present these results because it is known that there are more efficient tests for location and scale alternatives than the (Kolmogorov-)Smirnov test (Büning and Trenkler, 1994, p. 124). However, we note that Büning (2002) introduced a modified (Kolmogorov-) Smirnov test for location and scale alternatives that can be applied in the case of extremely right-skewed distributions.

There are several further location-scale tests that are based on the sum of two test statistics. For instance, Pettitt (1976) combined the WMW test with the Mood test. Murakami (2007) proposed to combine the $\tilde{B}^*$ test with the Mood test. Neuhäuser et al. (2011) combined two WMW statistics; however, one of them is computed after a Levene transformation (see Section 3.4).

Instead of calculating the sum of two test statistics, one can use the maximum for the construction of a location-scale test. Lepage (1971) already mentioned that possibility but did not investigate it. The Lepage test with the statistic L is usually more powerful than an analogous maximum test when there are differences in both location and variability. The maximum test is more powerful when there is one type of difference only. The reason is that differences in location and variability can accumulate in L, but not in a maximum. In a more general context, Cox (1977) recommended a maximum test when only a pure deviation from the null hypothesis is possible. The test based on the sum or a similar linear combination can be more powerful in the case of a mixed deviation from the null hypothesis. This holds in particular when combining more than two tests (see, e.g., Neuhäuser, 2003a).

According to Cox (1977, pp. 54–55), a maximum test has "useful diagnostic properties" and therefore a "greater general informativeness," as the maximal test statistic indicates the type of the deviation from the null hypothesis (if any). With a sum statistic this is not directly possible, but the lower informativeness is compensated for by the fact that further tests can follow after a significant location-scale test. These further tests may detect specific deviations from the null hypotheses. It should be noted that the closed testing procedure as indicated above is not consonant; that is, the location-scale test can be significant, without any test of step 2 being significant. The coherence, however, is fulfilled. Because the two-step procedure is a closed test, a significance at step 2 is not possible without a significance in the location-scale test at step 1. If there is no significance at step 1, the procedure stops without performing any further test.

3.1.5   The Cucconi test

Cucconi (1968) introduced a location-scale test that is almost unknown, probably because it was published in Italian only. Recently, Marozzi (2008) pointed out this test and compared its power with that of the Lepage test. The test statistic of the Cucconi test is

$$C = \frac{U^2 + V^2 - 2\rho UV}{2(1 - \rho^2)},$$

where U, V, and ρ are defined as follows:

$$U = \frac{6\sum_{j=1}^{n_1} R_{1j}^2 - n_1(N+1)(2N+1)}{\sqrt{n_1 n_2(N+1)(2N+1)(8N+11)/5}},$$

$$V = \frac{6\sum_{j=1}^{n_1}(N+1-R_{1j})^2 - n_1(N+1)(2N+1)}{\sqrt{n_1 n_2(N+1)(2N+1)(8N+11)/5}},$$

$$\rho = \frac{2(N^2 - 4)}{(2N+1)(8N+11)} - 1.$$

Furthermore, R_{1j} denotes the rank of X_j, $j = 1, \ldots, n_1$. Analogously, R_{2j} denotes the rank of Y_j, $j = 1, \ldots, n_2$. In Section 2.3 a divergent notation with $R_1 < \cdots < R_{n_1}$ ($H_1 < \cdots < H_{n_2}$) was used. However, ranks in increasing order were used in Section 2.3.

For the data example of the colony sizes of bumblebees (see Table 3.2), the ranks R_{1j} are 3.5, 7, 8, 18, 2, 19, and 17. A test statistic value of $C = 1.6720$ results. For n_1, $n_2 \to \infty$ with $n_1/N \to \lambda \in\ ]0, 1[$, the negatively correlated statistics U and V are both asymptotically standard normal. Cucconi (1968) demonstrated that the asymptotic Cucconi test can reject the null hypothesis H_0^{LS} at a nominal level α if $C \geq -\ln(\alpha)$. If this asymptotic test is applied for the example data despite the small sample sizes, H_0^{LS} cannot be rejected at $\alpha = 5\%$ because $C = 1.6720 < -\ln(0.05) = 2.996$.

Obviously, a permutation test can be carried out with Cucconi's test statistic C. When analysing the example the exact test is based on all 50 388 permutations. The resulting p-value is $P_0(C \geq 1.6720) = 9\,696/50\,388 = 0.1924$. Hence, the non-significance of the asymptotic test is confirmed.

Marozzi (2008) compared the power of the tests proposed by Cucconi and Lepage in a simulation study. This comparison did not show a clear winner. In the presence of ties, the Cucconi test has a disadvantage. Without ties it makes no difference whether U and V are computed based on the data of the first or the second group. When the second instead of the first group is chosen to compute U and V, the two statistics have a different sign, but that is irrelevant for the squares U^2 and V^2 as well as the product UV, and therefore the test statistic C is unchanged. However, in the presence of ties, one obtains different values for C, depending on whether the observations of the first or the second groups are used to compute U and V. In the bumblebee example there is one tied group. When U and V are computed with the ranks from group 2, one obtains $C = 1.6876$ instead of 1.6720, the value mentioned above based on group 1. And with the test statistic, the p-value also changes; it is now 0.1888 instead of 0.1924.

3.1.6   Implementation in SAS

In order to apply the Cucconi test, the SAS macro presented in Section 2.3 can be used. Of course, the part of the program that computes the test statistic must be modified. To be precise, the following SAS code calculates the statistic C and can be inserted between the parts "Creation of all possible permutations" and "Definition of output":

```
/* Calculation of the Cucconi test statistic */
start test_sta(R1, N_total, n1, n2);
  sumU=(R1##2)[ ,+];
  sumV=((N_total+1-R1)##2)[ , +];
  Q=sqrt((n1#n2#(N_total+1)#(2#N_total+1)#(8#N_total+11))/5);
  U=(6#sumU-n1#(N_total+1)#(2#N_total+1))/Q;
  V=(6#sumV-n1#(N_total+1)#(2#N_total+1))/Q;
  p=((2#(N_total##2-4))/((2#N_total+1)#(8#N_total+11)))-1;
  C=(U##2+V##2-2#p#U#V)/(2#(1-p##2));

return (C);
finish;

/* Carrying out the tests */
Tab=REPEAT(T(ranks),P,1);

R1=choose(permutationen=0,.,Tab);
R1g=R1[loc(R1^=.)];
R1z=shape(R1g,P, n1);
Cr = test_sta(R1z, N_total, n1, n2);

test_st0=test_sta(T(ranks[1:n1]), N_total, n1, n2);
Pval=(Cr>=test_st0);
Pval=Pval[+]/P;
```

3.2   The Nonparametric Behrens–Fisher Problem

The problem to test for a difference in location, despite a possible difference in variability between the groups, is called the Behrens-Fisher problem. Because we do not assume that data are normally distributed, we face the nonparametric Behrens-Fisher problem.

In the nonparametric Behrens-Fisher problem we do not test the general alternative $F \neq G$. Instead, the aim is to show a tendency to smaller, or larger, values. Hence, the null hypothesis H_0^{BF}: $p = 1/2$ is tested against the alternative H_1^{BF}: $p \neq 1/2$ (Brunner and Munzel, 2002, p. 53), where p is the relative effect $p = P(X_i < Y_j) + 0.5P(X_i = Y_j)$ as already defined above. The distributions with $F = G$ are a subset of the distributions with $p = 1/2$. Therefore, it is appropriate to test H_0^{BF} versus H_1^{BF} in a second step, if a location-scale test could reject the null hypothesis $F = G$. Depending on the aim of the study, H_0^{BF} versus H_1^{BF} can also be tested directly without any preceding test.

To illustrate the meaning of the statement $p = 1/2$, we consider the parametric Behrens-Fisher problem as a special case of the nonparametric test problem H_0^{BF} versus H_1^{BF}. It is in fact a special case because for two independent normally distributed random variables $Z_1 \sim N(\mu_1, \sigma_1^2)$ and $Z_2 \sim N(\mu_2, \sigma_2^2)$, we have (Reiser and Guttman, 1986)

$$p = P(Z_1 < Z_2) = \Phi\left(\frac{\mu_2 - \mu_1}{\sqrt{\sigma_1^2 + \sigma_2^2}}\right).$$

Because $\Phi(0) = 0.5$, the hypothesis $\mu_1 = \mu_2$ is equivalent to $p = 1/2$, where the variances σ_1^2 and σ_2^2 may differ (see also Brunner and Munzel, 2000). Figure 3.4 shows the relationship between p and the difference of the means $\mu_2 - \mu_1$ of two normally distributed random variables with variance 1.

Bernoulli variables can also be used to illustrate the relative effect (Brunner and Munzel, 2002, p. 23). Two independent random variables X and Y have the probabilities of success q_1 and q_2, respectively; that is, $P(X = 1) = q_1$ and $P(Y = 1) = q_2$. Hence, $P(X = 0) = 1 - q_1$ and $P(Y = 0) = 1 - q_2$. We have $X < Y$ if $X = 0$ and $Y = 1$; this happens with probability $(1 - q_1)q_2$. The equality $X = Y$ holds if X and Y are both 0, or if both are 1. The corresponding probabilities are $(1-q_1)(1-q_2)$ and q_1q_2, respectively. It follows for the relative effect that

$$
\begin{aligned}
p = P(X < Y) + 0.5P(X = Y) &= (1 - q_1)q_2 + 0.5[(1 - q_1)(1 - q_2) + q_1q_2] \\
&= 0.5 + 0.5(q_2 - q_1).
\end{aligned}
$$

Thus, the relative effect p is $1/2$ if and only if $q_1 = q_2$.

In general, without making any assumptions about the shape of the distributions, we can conclude that the observations in group 1 (X values) tend to be smaller in comparison to those of group 2 (Y values) if $p > 0.5$. In the case of $p < 0.5$, we conclude that the observations in group 1 tend to be larger than those of group 2. Under the null hypothesis H_0^{BF}: $p = 0.5$, neither group generally has larger values than the other and $P(X > Y) = P(X < Y)$ holds (see, e.g., Delaney and Vargha, 2002; Neuhäuser and Ruxton, 2009b).

Thus, the relative effect p is suitable to test H_0^{BF} versus H_1^{BF} in the nonparametric Behrens-Fisher problems. However, the two distributions F and G might differ even under the null hypothesis. Therefore, one cannot perform a permutation test in the usual way as described above (see, e.g., Romano, 1990; Brunner and Munzel, 2002, p. 75; Huang et al., 2006).

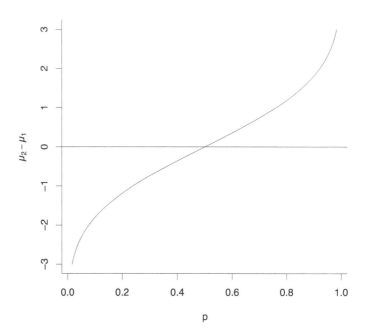

FIGURE 3.4: The relationship between the relative effect p and the difference of the means $\mu_2 - \mu_1$ of two normal distributions with variance 1.

Fligner and Policello (1981) introduced an asymptotic test that can be regarded as a modification of the WMW test for the test problem H_0^{BF} versus H_1^{BF}. Brunner and Munzel (2000) further generalized this test so that arbitrary distribution functions are acceptable, with the exception of a one-point distribution; that is, all data points have the same value. Moreover, they gave a small sample approximation.

The test statistic suggested by Brunner and Munzel (2000; see also Brunner and Munzel, 2002, Section 2.1.3) is defined as

$$W_{BF} = \sqrt{\frac{n_1 n_2}{N}} \frac{\bar{R}_2 - \bar{R}_1}{\hat{\sigma}_{BF}},$$

where $\bar{R}_i$ denotes the arithmetic mean of the ranks of group i; moreover,

$$\hat{\sigma}_{BF}^2 = \sum_{i=1}^{2} \frac{N \tilde{S}_i^2}{N - n_i} \quad \text{and} \quad \tilde{S}_i^2 = \frac{1}{n_i - 1} \sum_{j=1}^{n_i} \left(R_{ij} - R_{ij}^{(i)} - \bar{R}_i + \frac{n_i + 1}{2} \right)^2,$$

where the R_{ij}, $i = 1, 2$, $j = 1, \ldots, n_i$, denote the combined-samples ranks as above: R_{1j} is the rank of X_j, and R_{2j} the rank of Y_j; $R_{ij}^{(i)}$ is the (within) rank of the respective observation; that is, the rank among the n_i observations within group i. Note that

$$\hat{p} = \frac{1}{N} \left(\bar{R}_2 - \bar{R}_1 \right) + 0.5$$

is an unbiased and consistent estimator for p (Brunner and Munzel, 2000).

The test statistic W_{BF} is asymptotically standard normal (Brunner and Munzel, 2000). As with the t test for the parametric Behrens-Fisher problem, an approximation with a t distribution can be used in the case of small samples. To be precise, the t distribution with

$$\mathrm{df} = \frac{\left(\sum\limits_{i=1}^{2} \frac{\tilde{S}_i^2}{N - n_i} \right)^2}{\sum\limits_{i=1}^{2} \frac{\left(\tilde{S}_i^2 / (N - n_i) \right)^2}{n_i - 1}}$$

can be used. When there are no ties, this approximation is acceptable if $\min(n_1, n_2) \geq 10$; the standard normal distribution can be used if $\min(n_1, n_2) \geq 20$ (Brunner and Munzel, 2002, p. 80).

There are further methods to test the null hypothesis H_0^{BF}: $p = 1/2$ (see, e.g., Delaney and Vargha, 2002). Compared to alternative tests, the Brunner-Munzel test works well (Delaney and Vargha, 2002; Neuhäuser and Lam, 2004; Reiczigel et al., 2005), although a general recommendation to use the Brunner-Munzel test might be too simplistic. We refer the reader to Fagerland and Sandvik (2009a, 2009b), who compared the Brunner-Munzel tests with other tests within a wide ranging simulation study.

When performing the Brunner-Munzel test, at least ten observations per group are needed to apply the small sample approximation using the t distribution proposed by Brunner and Munzel (2000), as mentioned above. In the study performed by Reiczigel et al. (2005), the Brunner-Munzel test showed up well for samples sizes of at least thirty. Hence, a permutation test for the nonparametric Behrens-Fisher problem would be desirable, in particular for discrete distributions for which the rules of thumb mentioned in the previous paragraph cannot be generally used. The reason is that the appropriateness of the approximation depends on the number and pattern of ties (Brunner and Munzel, 2002, p. 80).

3.2.1 A Permutation Test with the Brunner–Munzel Statistic

As already mentioned, a usual permutation test is not possible because the two distributions F and G may differ under the null hypothesis H_0^{BF}. However, Janssen (1997) demonstrated that a permutation test with a studentized test statistic can asymtotically guarantee the nominal significance level if the two potentially unequal variances exist and if some requirements are fulfilled

(Janssen, 1997). Based on this result, Neubert and Brunner (2007) suggested a permutation test with the Brunner-Munzel test statistic W_{BF}. When this permutation test is carried out, $\hat{\sigma}_{BF}$ must be computed for each permutation because $\hat{\sigma}_{BF}$ can take on different values for different permutations. Thus, the entire statistic W_{BF}—as defined above—has to be calculated for each permutation.

When the smallest value in one group is larger than all values of the other group, we have $\hat{\sigma}^2_{BF} = 0$ as the estimated variance. Then, the test statistic W_{BF} cannot be calculated. Such a situation might be unlikely, although it can occur in particular when there are small samples and large differences between the groups. However, when performing a permutation test, this situation always occurs for the two extreme permutations that assign all small values into one group. Neubert and Brunner (2007) suggested $\hat{\sigma}^2_{BF} = N/(2n_1n_2)$ in such a situation in which we would have $\hat{\sigma}^2_{BF} = 0$ when applying the formula given above.

Simulation studies showed that the permutation test with W_{BF} has an acceptable actual level even for small and unbalanced sample sizes (Neubert and Brunner, 2007; Neuhäuser and Ruxton, 2009b). The power of the test was also acceptable compared to alternative methods (Neubert and Brunner, 2007). However, Fagerland et al. (2011) presented simulation results that indicate that the permutation test does not control the nominal significance level in case of discrete numerical data with few possible values, such as 0, 1, and 2. The simulated actual level was as large as 0.098 for $\alpha = 0.05$.

At www.ams.uni-goettingen.de/de/sof/, Brunner and Neubert (2007) offer a macro written in SAS/IML that can perform the Brunner-Munzel test both as an approximate permutation test and as an asymptotic test based on the t distribution. If an exact permutation test—that is, a test using all possible permutations—is required, the SAS/IML program given in Section 2.3 can be used in order to generate the permutations. An R program can be found at www.biostat.uni-hannover.de/staff/neuhaus/BMpermutation_test.txt (Neuhäuser and Ruxton, 2009b).

3.2.2   Example

The Brunner-Munzel test can be applied to the example data on bumblebee colonies given in Table 3.2. As mentioned, the ranks of group 1 are 3.5, 7, 8, 18, 2, 19, and 17. Hence, the mean of these ranks is $\bar{R}_1 = 10.643$. For group 2 we have $\bar{R}_2 = 9.625$. The estimated variance is $\hat{\sigma}^2_{BF} = 46.986$; thus the test statistic W_{BF} is equal to -0.3122. The two-sided p-value of the exact permutation test based on all 50,388 permutation is $P_0(|W_{BF}| \geq 0.3122) = 37,591/50,388 = 0.7460$.

Despite the small sample sizes, we give the result of the approximate test using the t distribution for illustrative purposes. Because df $= 7.52$, the two-sided p-value is 0.7633.

3.3 Bootstrap Tests

For any permutation the observations are allocated to two (or more) groups. Each observed value is allocated to exactly one group. Thus, observations are drawn "without replacement" to form a permutation. In contrast, for a bootstrap test the observations are drawn "with replacement." As a consequence, some values might be used more than once, and some other values are not allocated to any group.

When generating all the possible permutations for an exact permutation test, one must generate a total of $\binom{N}{n_1}$ permutations. Because the draw is with replacement, there are more possibilities for the bootstrap, namely, $\binom{N+n_1-1}{n_1}\binom{N+n_2-1}{n_2}$ different bootstrap samples if all N observations of the two samples are different. These possible bootstrap samples, however, are not equiprobable under the null hypothesis (Efron and Tibshirani, 1993, p. 58). For example, in the case of $n_1 = 3$, $n_2 = 5$ (hence $N = 8$), and no ties, there are 56 different permutations, but 95,040 different bootstrap samples.

The bootstrap null distributions may be less discrete than the permutation null distribution because of this difference between the number of possible permutations and bootstrap samples. As a consequence, when sample sizes are very small, the bootstrap test can be less conservative than the corresponding permutation test (Neuhäuser and Jöckel, 2006). However, when the sample sizes are not very small, it is usually not possible to consider all potential bootstrap samples. Instead, a simple random sample of B bootstrap samples is drawn. The number B should be at least 1,000 (Rózsa et al., 2000). When the samples are small enough that all possible permutations can be considered, but large enough that not all potential bootstrap samples can be drawn, the permutation test has the advantage of being exact and not approximate (Efron and Tibshirani, 1993, p. 216; see also Romano, 1989; Janssen and Pauls, 2003).

In lieu of the permutation tests discussed in Chapter 2, bootstrap tests are also possible using the same test statistics. However, the differences between the permutation and bootstrap tests are usually small (Efron and Tibshirani, 1993, p. 216). And the permutation tests have the already-mentioned benefit that they are exact and not approximate. On the other hand, bootstrap tests have a broader area of application. Before this is discussed in more detail, some more particulars of the bootstrap test will be explained.

When performing the Fisher-Pitman permutation test, it does not make a difference whether the student's t statistic or simply the numerator of this statistic, that is, the difference in means, is used as the test statistic. When performing a bootstrap test it makes a difference: Some observations are included twice or multiple times and some other observations are not included at all in the bootstrap sample; therefore, the variance estimation changes. Thus, the denominator of the t statistic differs between different bootstrap samples; hence, the denominator cannot be omitted without consequences. In this case it is recommended to use the studentized statistic, that is, the entire t statistic, as the test statistic for the bootstrap test (Hall and Wilson, 1991; Efron and Tibshirani, 1993, p. 221).

Therefore, the two-sided **bootstrap t test** can be carried out as follows:

(a) Calculate the test statistic $t = \dfrac{\bar{X} - \bar{Y}}{S \cdot \sqrt{\frac{1}{n_1} + \frac{1}{n_2}}}$ for the actually observed values.

(b) Draw $n_1 + n_2$ values with replacement from the observations X_1, ..., X_{n_1}, Y_1, ..., Y_{n_2}; the first n_1 drawn values are assigned to group 1, and the further n_2 values to group 2. Compute the test statistic t for this bootstrap sample.

(c) Repeat step (b) B times.

(d) Compute the p-value as the proportion of bootstrap samples for which the absolute value of the test statistic—that is, $|t|$—is at least as large as $|t|$ for the originally observed values (the latter is calculated in step a).

It is often proposed to transform the observations X_1, ..., X_{n_1}, Y_1, ..., Y_{n_2} prior to draw the bootstrap samples (Hall and Wilson, 1991; Manly, 2007). To be precise, the group means $\bar{X}$ and $\bar{Y}$, respectively, are subtracted in order to get a common mean for both groups. Manly's (2007) rationale for this transformation is that the null hypothesis is made to be true for the bootstrap. When applying this transformation for the test defined above one has to compute the transformations $\dot{X}_i = X_i - \bar{X}$ and $\dot{Y}_j = Y_j - \bar{Y}$ after step (a). Then, the bootstrap sample has to be drawn in step (b) from the transformed values. If a test statistic based on ranks is used, a transformation using medians might be appropriate (Reiczigel et al., 2005).

3.3.1   The SAS Procedure MULTTEST

The SAS procedure PROC MULTTEST is—as indicated by its name—designed for multiple test problems. However, one can also use this procedure to carry out the Fisher-Pitman permutation test (FPP test, see Section 2.1) as well as the bootstrap t test defined above (Westfall and Soper, 1994).

The FPP test can be invoked in the following way (**group** and **count** are again variable names):

```
PROC MULTTEST PERMUTATION DATA=example1;
   CLASS group;
   TEST MEAN(count);
   CONTRAST "1 vs. 2" -1 1;
RUN;
```

Based on the example data of Good (2001, p. 56), this program generates the following output:

```
The Multtest Procedure

Model Information
```

Test for continuous variables	Mean t-test
Tails for continuous tests	Two-tailed
Strata weights	None
P-value adjustment	Permutation
Center continuous variables	No
Number of resamples	20000
Seed	148203001

```
Contrast Coefficients
```

	group	
Contrast	1	2
1 vs. 2	-1	1

```
Continuous Variable Tabulations
```

Variable	group	NumObs	Mean	Standard Deviation
count	1	4	109.7500	13.9613
count	2	4	40.7500	37.2682

```
p-Values
```

Variable	Contrast	Raw	Permutation
count	1 vs. 2	0.0133	0.0576

An approximate permutation test is carried out here. Nevertheless, no confidence interval for the *p*-value is given, in contrast to the procedure NPAR1WAY. The option PERMUTATION within the PROC MULTTEST statement selects the permutation test. With the option BOOTSTRAP, a bootstrap test can be performed:

```
PROC MULTTEST BOOTSTRAP DATA=example1;
  CLASS group;
  TEST MEAN(count);
  CONTRAST "1 vs. 2" -1 1;
RUN;
```

Then, the p-value of the bootstrap test is listed in the output:

```
                        p-Values


   Variable    Contrast              Raw       Bootstrap


    count      1 vs. 2              0.0133      0.0097
```

The big difference between the permutation and the bootstrap p-values is striking. The much smaller p-value of the bootstrap test is caused by the less discrete null distribution of the bootstrap in comparison with the permutation distribution. As discussed, the difference in the discreteness can occur in the case of very small samples such as in this example with $n_1 = n_2 = 4$.

There are some further useful options within the PROC MULTTEST statement. The option CENTER invokes the transformation mentioned above, that is, the observations are transformed by subtracting the group means. By default, SAS performs this transformation for a bootstrap test, but waives it for a permutation test. When a bootstrap test without this transformation is desired, one can do so with the option NOCENTER. The option N= or NSAMPLE= defines the number of considered permutations or bootstrap samples, respectively. By default, NSAMPLE = 20,000. It is not possible to consider all possible permutations or all potential bootstrap samples. Hence, with PROC MULTTEST the FPP test is only possible as an approximate test. As a consequence, one should prefer the procedure NPAR1WAY when the samples are small and all permutations should be considered.

A further option is S= or SEED= to specify the seed value. The option OUTSAMP=SAS-data-set generates an SAS data set that includes the chosen permutations or bootstrap samples, respectively. Obviously, this data set can be very huge, in particular when a large value for NSAMPLE is chosen.

Within the TEST statement one must specify MEAN in order to perform a t test. In this statement the options LOWERTAILED or UPPERTAILED are possible in order to carry out a one-sided test. In the CONTRAST statement the two values -1 and 1 are used for a two-sample comparison. The following program performs a one-sided test:

```
PROC MULTTEST BOOTSTRAP DATA=example1;
  CLASS group;
  TEST MEAN(count/LOWERTAILED);
  CONTRAST "1 vs. 2" -1 1;
RUN;
```

It should be noted that for the Peto test as well as the Armitage test (see Chapter 5), exact permutation tests are possible with PROC MULTTEST. Moreover, Fisher's exact test is also possible with PROC MULTTEST.

When a bootstrap test based on another test statistic should be programmed in SAS, the option OUTSAMP=SAS-data-set of the procedure MULTTEST can be applied to write the chosen bootstrap samples into an SAS data set. This data set can then be used to carry out the bootstrap test based on another test statistic that is not implemented in PROC MULTTEST. Obviously, the option OUTSAMP= can also be used together with the option PERMUTATION in the PROC MULTTEST statement in order to perform an approximate permutation test with the chosen permutations.

Alternatively, the selection of bootstrap samples can be programmed within SAS/IML or within a data step. If the latter, one can use the SAS function RANTBL to select values for the bootstrap sample. An SAS/IML program proposed by Good (2001, p. 204) is based on the SAS function RANUNI; here again, Y denotes the vector of data and Ystar the generated bootstrap sample:

```
proc iml;
    Y={11, 13, 10, 15, 12, 45, 67, 89};
    n=nrow(Y);
    U=ranuni(J(n,1, 3571));     *3571 is the seed;
    I=int(n*U + J(n,1,1));
    Ystar=Y(|I,|);
    print Ystar;
quit;
```

3.3.2 A Bootstrap Test for the Behrens–Fisher Problem

Bootstrap tests have a broad area of application and can also be applied to testing the location problem without assuming homoscedasticity, that is, equality of variances. For this purpose, a bootstrap test can be performed with the t statistic modified for the Behrens-Fisher problem (Efron and Tibshirani, 1993, pp. 222–223):

$$t_{BF} = \frac{\bar{X} - \bar{Y}}{\sqrt{\frac{S_1^2}{n_1} + \frac{S_2^2}{n_2}}},$$

where $S_1^2 = \frac{1}{n_1-1}\sum_{i=1}^{n_1}(X_i - \bar{X})^2$ and $S_2^2 = \frac{1}{n_2-1}\sum_{j=1}^{n_2}(Y_j - \bar{Y})^2$. The null distribution of the statistic $|t_{BF}|$, needed for the two-sided test, can be estimated using the bootstrap in the following way (cf. Efron and Tibshirani, 1993):

(a) Calculate $|t_{BF}|$ for the actually observed values.

(b) Transform the observed values in order to have a common mean in both groups: $\dot{X}_i = X_i - \bar{X}$ and $\dot{Y}_j = Y_j - \bar{Y}$.

(c) Draw two separate bootstrap samples of size n_1 and n_2 from the transformed values $\dot{X}_i$ ($i = 1, \ldots, n_1$) and $\dot{Y}_j$ ($j = 1, \ldots, n_2$), respectively, with replacement. Calculate $|t_{BF}|$ for this bootstrap sample.

(d) Repeat step (c) B times.

The p-value is then the proportion of bootstrap samples for which $|t_{BF}|$ is at least as large as $|t_{BF}|$ based on the actual observations. When performing a one-sided test, the test statistic is used without the absolute value, that is, t_{BF}. In the case of small samples, especially in combination with ties, it can occur that all values in a bootstrap sample are identical (cf. Westfall and Young, 1993, p. 91). When this happens in both groups, the statistic t_{BF} cannot be computed. In such a case with $\max(S_1^2, S_2^2) = 0$, we set $t_{BF} = \infty$ or $-\infty$, depending on the sign of the numerator.

The distributions F and G may differ under the null hypothesis. Therefore, separate bootstrap samples of size n_1 and n_2, respectively, are drawn in step (c). This approach is called "separate-sample bootstrap" (McArdle and Anderson, 2004). An analogous procedure is not possible when applying a permutation test because the draw is without replacement and, consequently, a "separate-sample" permutation test can generate no other permutation than the one corresponding to the original observations.

The "separate-sample bootstrap" t test cannot be carried out directly with the SAS procedure MULTTEST. However, one can use this procedure to generate the two separate bootstrap samples. The following SAS program performs the bootstrap t test with the test statistic t_{BF} for the example data on bumblebee colonies (here with $B = 20{,}000$):

```
DATA colonies;
  INPUT group count dummy @@;
CARDS;
1 14 1 1 24 1 1 26 1 1 98 1 1 12 2 1 105 2 1 85 2
2 40 1 2 14 1 2 18 1 2 28 1 2 11 1 2 39 2 2 17 2
2 37 2 2 52 2 2 30 2 2 65 2 2 35 2
;
RUN;

PROC SORT;
  BY group;
RUN;

PROC MEANS MEAN NOPRINT;
  BY group;
  VAR count;
```

```
  OUTPUT OUT=m MEAN=meanw;
RUN;

DATA _null_;
  SET m;
  IF group=1 THEN DO;
   CALL SYMPUT ('meanw_1',meanw);
  END;
  IF group=2 THEN DO;
   CALL SYMPUT ('meanw_2',meanw);
  END;
RUN;

DATA colonies;
  SET colonies;
  IF group=1 THEN DO;
   trans_x=count-&meanw_1;
  END;
  IF group=2 THEN DO;
   trans_x=count-&meanw_2;
  END;
RUN;

PROC MULTTEST DATA=colonies
  BOOTSTRAP NOCENTER N=20000 OUTSAMP=p1 NOPRINT;
  CLASS dummy;
  TEST MEAN(trans_x);
  CONTRAST "1 vs. 2" -1 1;
  STRATA group;
RUN;

ODS OUTPUT TTests=s3;
ODS LISTING CLOSE;
PROC TTEST DATA=p1;
  BY _sample_;
  CLASS _stratum_;
  VAR trans_x;
RUN;
ODS LISTING;

*Calculation of the bootstrap p-value;
DATA s3;
  SET s3;
```

```
   WHERE METHOD="Satterthwaite";
RUN;

ODS OUTPUT TTests=origvalue;
PROC TTEST DATA=colonies;
   CLASS group;
   VAR count;
RUN;

DATA _null_;
   SET origvalue;
   WHERE METHOD="Satterthwaite";
   CALL SYMPUT ('t_orig',tValue);
RUN;

DATA s3;
   SET s3;
   t_orig=&t_orig;
RUN;

DATA all4;
   SET s3;
   p_value=(ABS(tValue)>=ABS(t_orig))/20000;
RUN;

PROC MEANS SUM DATA=all4;
   VAR p_value;
RUN;
```

The variable called **dummy** is needed so that the procedure MULTTEST can carry out two separate two-sample bootstrap tests, one for each group. The results of these tests are not wanted; therefore, the option NOPRINT is used in the PROC MULTTEST statement. As a result, the test results do not occur in the output window. In contrast to the test results, the selected bootstrap samples are required; these samples are written into a SAS data set using the option OUTSAMP.

For the SAS procedure TTEST there is no option NOPRINT. With the command **ODS OUTPUT TTests=s3;** the results of the t tests are written into a SAS data set called **s3**. Using the example data on colony sizes and $B = 20{,}000$ we get $t_{BF} = 1.20$ and a p-value of 0.2548. When applying the Welch t test, the resulting p-value is slightly larger: 0.2679.

The SAS procedure SURVEYSELECT offers a further possibility to draw bootstrap samples:

```
PROC SURVEYSELECT DATA=example1
```

```
METHOD=URS OUT=bootst1 REP=1000 N=4 SEED=34567 OUTHITS;
  STRATA group;
RUN;
```

The method URS requests a simple random sample with replacement as needed for the bootstrap samples. With the command OUT=, one can term the SAS data set that includes the bootstrap samples. The number of bootstrap samples is specified using REP=. At N=, one must provide the sample size per group, and SEED= offers the option to specify the seed value for the random number generation. The option OUTHITS is required so that values that occur more than once in a bootstrap sample are included as a separate observation for each selection of that observation. By default, the output data set would contain only one observation for each selected value, even if it is selected more than once into the bootstrap sample.

The STRATA statement specified the variable **group** in order to draw separate bootstrap samples in this example. In the case of unbalanced sample sizes of the two groups one must specify both numbers. For instance, for $n_1 = 4$ and $n_2 = 5$, the corresponding option must be

N=(4,5)

The bootstrap test for the Behrens-Fisher problem outlined here can be modified. Reiczigel et al. (2005) investigated a modification that is suitable when the data clearly deviate from a normal distribution. The transformation in step (b) is performed using medians, and the test statistic t_{BF} is computed based on ranks rather than the original values; see Reiczigel et al. (2005) for further details.

3.3.3 The D.O Test, A Combination of Permutation Test and Bootstrap

Manly (1995) as well as Manly and Francis (1999) proposed an approach for how a permutation test can be applied to test for a difference in location in case of potentially unequal variances. This so-called D.O test was used by Manly and Francis (2002) in a two-step procedure as the test for the location problem at step 2. The D.O test is also suitable for the comparison of more than two samples. However, here we consider the two-sample situation only. The test consists of the following five steps and a bootstrap calibration that is described later.

(a) Calculate the FPP test statistic P for the original observations.

(b) Solve the equations

$$\hat{\mu} = \frac{n_1 \bar{X}/\hat{\sigma}_1 + n_2 \bar{Y}/\hat{\sigma}_2}{\sum_{i=1}^{2}(n_i/\hat{\sigma}_i)} \quad \text{and}$$

$$\hat{\sigma}_1^2 = \frac{\sum_{j=1}^{n_1}(X_j - \hat{\mu})^2}{n_1} , \ \hat{\sigma}_2^2 = \frac{\sum_{j=1}^{n_2}(Y_j - \hat{\mu})^2}{n_2}$$

iteratively.

(c) Determine

$$U_{1j} = \hat{\mu} + \frac{(X_j - \hat{\mu})}{\hat{\sigma}_1} , \ j = 1, \ldots, n_1, \text{ and}$$

$$U_{2j} = \hat{\mu} + \frac{(Y_j - \hat{\mu})}{\hat{\sigma}_2} , \ j = 1, \ldots, n_2,$$

using the estimates found in step (b).

(d) Randomly reallocate the u values calculated in step (c) to two groups of sizes n_1 and n_2, respectively. The corresponding set of unadjusted values is computed by the back-transformation $x = \hat{\mu} + \hat{\sigma}_i(u - \hat{\mu})$. Calculate the FPP test statistic P for this set of unadjusted values.

(e) Repeat step (d) $M - 1$ times.

The p-value of the test is then calculated based on the resulting permutation null distribution of the test statistic P. This permutation null distribution is determined upon the $M - 1$ permutations of step (e) and the original observations. Obviously, this is an approximate permutation test. Of course, in the case of small samples, all possible permutations can be considered in step (e).

Due to the chosen transformation, the adjusted u values have, under the null hypothesis H_0^{BF}, identical expected values and identical variances in both groups. The common variance is 1. The u values and not the actual observations are used in step (d), where the values are randomly allocated to two groups of sizes n_1 and n_2. Thus, values with a homogeneous variance are permuted. In this way, the potential heteroscedasticity of the original observations is circumvented.

According to the simulation studies of Francis and Manly (2001), this D.O test is α robust, that is robust with regard to size, in the case of unequal variances if the distributions do not differ much from a normal distribution. However, the test can have excessive size with data from distributions with very high skewness and/or kurtosis. Therefore, Francis and Manly (2001) introduced a bootstrap calibration. This calibration has the aim of calibrating the test, if necessary, by changing the significance criterion of the D.O test so that the nominal and actual sizes agree. An adjustment in the significance criterion means that the p-value is compared with a bound other than the nominal level α.

However, such a change in the significance criterion was abandoned by Manly and Francis (2002). The bootstrap is still used, but only to determine whether or not the result of the D.O test is reliable. Hence, rather than try to make the D.O test more reliable by calibration, it is investigated whether or not the result is reliable; that, is whether or not the nominal level is significantly exceeded. If not, the result of the D.O test is accepted. If the D.O test is not reliable, the null hypothesis cannot be rejected, irrespective of how small the p-value of the D.O test might be. Obviously, if the D.O test is not significant, the bootstrap calibration is not necessary.

The bootstrap procedure consists of the following steps (Francis and Manly, 2001, p. 718). Without loss of generality, we assume $n_1 \geq n_2$. The true, but unknown population variances are denoted by σ_1^2 and σ_2^2.

(a) The chosen test is performed based on the original observations.

(b) The observations of group 1 are used as the bootstrap distribution to be resampled with replacement. This distribution has the expected value $\bar{X}$ and the variance V_1, that is, the empirical variance of group 1, $V_1 = \sum_{j=1}^{n_1} (X_j - \bar{X})^2 / n_1$.

(c) The bootstrap distribution is sampled to create bootstrap samples of the sizes n_1 and n_2. The variances within these samples, V_{B1} and V_{B2}, are computed, this time with the denominators $n_1 - 1$ and $n_2 - 1$, respectively.

(d) The ratio $R_i = V_{Bi}/V_1$ can be regarded as approximately a random variable from the distribution of the ratio sample variance/true variance for a sample of size n_i from the source distribution. Thus

$$ R_i \approx \frac{S_i^2}{\sigma_i^2} \, , $$

implying that $\sigma_{Bi}^2 = S_i^2 V_1 / V_{Bi}$ is a plausible value for the population variance of group i.

(e) The bootstrap distribution is then sampled again with replacement to get new bootstrap samples of sizes n_1 and n_2. These samples are adjusted so that the variance in sample i is equal to σ_{Bi}^2, but the central locations are unchanged. The values within the bootstrap sample are denoted by X_{Bij}, $i = 1, 2$, $j = 1, \ldots, n_i$. The adjustment is done by transforming X_{Bij} to $\bar{X} + (X_{B1j} - \bar{X})\sigma_{B1}/V_1$ and $\bar{Y} + (X_{B2j} - \bar{Y})\sigma_{B2}/V_1$, respectively.

(f) The test already performed in step (a) is now carried out on the new adjusted bootstrap sample.

(g) The steps (c) through (f) are repeated $\tilde{M}$ times. The proportion of significant results in step (f) is determined. If this proportion significantly exceeds α—to be precise, if the proportion is larger than $\alpha + 1.64 \sqrt{\alpha(1-\alpha)/\tilde{M}}$—the test is regarded as not reliable, and a possibly found significance in step (a) is not accepted.

Francis and Manly (2001) also investigated a bootstrap that is not only based on one sample (cf. step (b)). That approach, however, was not advantageous, for details see Francis and Manly (2001, p. 719). Thus, only the data of one sample are used. However, the larger sample is always used.

In total, three results are possible when applying the approach proposed by Manly and Francis (2002):

1. The D.O test is not significant.

2. The D.O test is significant, but is considered unreliable based on the bootstrap calibration.

3. The D.O test is significant and reliable.

The null hypothesis can be rejected in the latter case only. Examples of applications of the D.O test can be found in Chapter 7.

Instead of the FPP test, other tests can also be used in the D.O procedure, including the bootstrap calibration. In particular, rank tests are possible (Neuhäuser and Manly, 2004).

However, independent of the chosen test statistic, there are two objections against a D.O test. On the one hand, the D.O test procedure is very complex. In addition to the usual procedure of a permutation test, steps such as the estimation of mean and variances, the transformations between u and original values, as well as the bootstrap validation are necessary. The latter is not only complex, but also reduces the power. This loss of power is inevitable because the bootstrap is required for compliance with the nominal level. Manly and Francis (2002) presented results for a Johnson distribution with skewness 5.2 and kurtosis 40.3 (for a detailed definition of this distribution, see Manly and Francis, 2002): For this distribution, the power of the D.O test with the bootstrap validation was never larger than 60% although the power without the bootstrap could be nearly 100%. For normally distributed data, the power is also reduced by the bootstrap validation, but a power of more than 60% is still possible (Manly and Francis, 2002, p. 645).

3.3.4   Which Test Can Be Recommended for the Behrens–Fisher Problem?

The D.O test cannot be recommended because of its complexity and the loss of power due to the bootstrap validation. Sometimes, a procedure called "rank Welch test," where the statistic t_{BF} is computed using the ranks, is applied. However, this test can be very anticonservative. In the simulations presented by Delaney and Vargha (2002), actual levels of more than 0.15 occurred in case of $\alpha = 0.05$. Thus, this test cannot be recommended, either.

The Brunner-Munzel test as well as the bootstrap test with the statistic t_{BF} remain. When applying the Brunner-Munzel test, the null hypothesis

TABLE 3.4: The simulated size of the bootstrap t_{BF} test as well as the Brunner–Munzel test with W_{BF}, both as an approximate test based on the t distribution and as a permutation test, for different symmetric distributions

Test	Uniform	Normal	t (3 df)
Bootstrap test with t_{BF}	0.046	0.044	0.033
Approximate W_{BF} test	0.056	0.053	0.054
W_{BF} permutation test	0.049	0.047	0.049

H_0^{BF}: $p = 1/2$ is tested. The statistic t_{BF} compares the means of the two groups. In the case of asymmetric distributions, the two null hypotheses H_0^{BF} and the equality of population means differ. Thus, one cannot directly compare the two tests. For a robust comparison of means, the bootstrap test based on t_{BF} can be proposed. When testing the null hypothesis $p = 1/2$, the Brunner-Munzel test is suitable. Depending on the sample sizes and the occurrence of ties, this test might be performed as a permutation test. However, the question remains as to which test can be suggested for symmetric distributions.

Table 3.4 displays the actual size of the bootstrap t_{BF} test as well as the Brunner-Munzel test with W_{BF}. Because the sample sizes are $n_1 = n_2 = 10$, the approximation using the t distribution can be used for the latter test, as suggested by Brunner and Munzel (2000). In addition, a permutation test based on W_{BF} was applied. For the bootstrap test $M = 1,000$ bootstrap samples are drawn each time.

The size of the approximate W_{BF} test often exceeds the nominal level; however, the deviation is small. As in the simulation results presented by Brunner and Munzel (2000), the size is not larger than 5.7%. Thus, the size can be regarded as acceptable. According to Brunner and Munzel's (2000) simulations, this result also holds for larger samples. Results for unbalanced samples can be found in Table 3.5. In this situation the tests investigated here have an acceptable size, even in the case when the larger group has the smaller variance—in contrast to the tests FPP, WMW, and BWS (cf. Table 3.1). However, the permutation test with W_{BF} is slightly anticonservative in this situation.

The simulated power is displayed in Table 3.6. For normally distributed data there is hardly a difference in power among the tests. This also holds in the case of heterogeneous variances (Neuhäuser, 2003b). For other distributions there is no clear winner. However, there can be larger differences, especially in cases where the W_{BF} test is more powerful (Neuhäuser, 2003b). The W_{BF} test can also be more powerful when there is no difference in size. For example, for a t distribution with three degrees of freedom, the simulated size of the bootstrap test is 0.033, the power for $\theta_2 = 2$ is 0.72. When performing the approximate W_{BF} test with a nominal level of 0.033 (instead of $\alpha = 0.05$), its power declines from 0.85 to 0.81; its actual level in this case is

TABLE 3.5: The simulated size of the bootstrap t_{BF} test as well as the Brunner–Munzel test with W_{BF}, both as an approximate test based on the t distribution and as a permutation test, for normally distributed data with mean 0 in both groups

			Standard Deviations in the Two Groups		
n_1	n_2	Test	1, 1	1, 2	1, 3
10	20	Bootstrap test with t_{BF}	0.048	0.046	0.050
		Approximate W_{BF} test	0.050	0.050	0.049
		W_{BF} permutation test	0.044	0.041	0.041
20	10	Bootstrap test with t_{BF}		0.052	0.045
		Approximate W_{BF} test		0.051	0.050
		W_{BF} permutation test		0.053	0.056

TABLE 3.6: The simulated power of the bootstrap t_{BF} test as well as the Brunner–Munzel test with W_{BF}, both as an approximate test based on the t distribution and as a permutation test, for different symmetric distributions

Test	Uniform $(\theta_2 = 0.4)$	Normal $(\theta_2 = 1.5)$	t (3 df) $(\theta_2 = 2)$
Bootstrap test with t_{BF}	0.83	0.88	0.72
Approximate W_{BF} test	0.80	0.88	0.85
W_{BF} permutation test	0.79	0.87	0.84

0.038. With a nominal level of 0.028, a simulated size of 0.033 results. Then, the power was simulated as 0.79, still more than 0.72, that is, the power of the bootstrap test (see Table 3.6).

Analogous results could be obtained for larger and unbalanced samples. Hence, the Brunner-Munzel test with the statistic W_{BF} might be recommended for the nonparametric Behrens-Fisher problem (see also Wilcox, 2003; Neuhäuser and Poulin, 2004; Rorden et al., 2007; Neuhäuser and Ruxton, 2009b; Neuhäuser, 2010). However, as mentioned above, the general recommendation to use the Brunner-Munzel test might be too simplistic (see also Fagerland et al., 2011). According to Fagerland and Sandvik (2009a, 2009b), the Welch t test, called the Welch U test by Fagerland and Sandvik, is preferable in some situations. As Fagerland and Sandvik (2009a) showed, the power comparison depends on a variety of settings including skewness, skewness heterogeneity, variance heterogeneity, sample size, and sample size ratio.

For asymmetric distributions the tests are associated with different null hypotheses. However, as Fagerland and Sandvik (2009a) also noted, the results are often interpreted as evidence for or against equality of means or medians. Because of the variety of methods, it might be important to identify situations in which most of the different methods give an identical result. Then one can confidently claim that values in one group are larger than in the other group. Here, we refer to Klar et al. (2010), who studied this approach very recently (see also Neuhäuser, 2011).

3.4 Tests for a Difference in Variability

A test for a difference in variability can be performed, as a test for the Behrens-Fisher problem (see Section 3.2), in a second step of a closed testing procedure after a significant location-scale test. The null hypothesis to be tested is H_0^S : $\theta_1 = 1$. There is no restriction for the location parameter θ_2. As mentioned above, we have $\theta_1 = \sigma_G/\sigma_F$ if the variances σ_F^2 and σ_G^2 of the distributions F and G, respectively, exist. Thus, in this case, $\theta_1 = 1$ is equivalent to equal variances. Under the alternative H_1^S we have $\theta_1 \neq 1$; that is, there is a difference between the two groups with regard to the scale parameter θ_1.

Of course, a test for a difference in variability can also be applied independently of a location or location-scale test when the primary question focuses on a difference in variability. This can be the case, for instance, in engineering, quality control, or in chemical applications (Pan, 2002).

Tests for a difference in variability designed for normally distributed data, as for example the F test, are not very robust. That means they are sensitive to deviations from normality (see, e.g., Box, 1953). Shoemaker (2003) suggested modifications of the F test that are more robust. However, these new tests can be anticonservative in the case of skewed data or heavy tails (Shoemaker, 2003).

A robust alternative to the F Test was introduced by Levene (1960). This test is based on the absolute values of the differences to the groups' means. To be precise, the transformed data

$$Z_{ij} = \begin{cases} |X_j - \bar{X}| & \text{if } i = 1 \ (j = 1, \dots, n_1) \\ |Y_j - \bar{Y}| & \text{if } i = 2 \ (j = 1, \dots, n_2) \end{cases}$$

are tested for a difference in location. Manly and Francis (2002) called this approach an indirect test for a difference in variability. The larger the variability in one group, the larger the values of Z_{ij} expected. Differences in variability are reflected in different means of the transformed variables Z_{ij}.

Furthermore, due to the transformation, the test is not sensitive to differences in location. Different population means cannot hide differences in variances. However, strictly speaking, the Levene test is a joint test of all even moments, but the effect of the variance predominates (van Valen, 2005, p. 32). The special situation of unbalanced sample sizes is discussed below in this chapter.

Levene (1960) applied parametric tests to the Z_{ij}. The Z_{ij} values, however, are not normally distributed, irrespective of whether or not the distributions F and G are normal. Therefore, nonparametric tests were suggested to analyze the transformed values Z_{ij} (Talwar and Gentle, 1977; Sokal and Braumann, 1980; Le, 1994; Manly and Francis, 2002). Moreover, the values are not independent; the Z_{ij}, $i = 1, 2$, $j = 1, \ldots, n_i$, are always correlated (see, e.g., O'Neill and Mathews, 2000). Nevertheless, the Levene test is an often used and powerful test (Schultz, 1983).

In place of the mean, other, more robust estimators of the central location can be used in Levene's transformation, for example, trimmed means or medians. According to Brown and Forsythe (1974), the absolute deviations from sample medians are preferable, in particular if the underlying distributions are asymmetric (see also Schultz, 1983; Büning, 2002). Because the distribution functions F and G can be asymmetric in the nonparametric model considered here, Levene's transformation is applied with medians. To be precise, the tests are based on

$$
\widetilde{Z}_{ij} = \begin{cases} \left| X_j - \widetilde{X} \right| & \text{if } i = 1 \ (j = 1, \ldots, n_1) \\[2ex] \left| Y_j - \widetilde{Y} \right| & \text{if } i = 2 \ (j = 1, \ldots, n_2), \end{cases}
$$

where $\widetilde{X}$ and $\widetilde{Y}$ denote the sample medians of the two groups. If n_1 and/or n_2 are even, the arithmetic means of the two central values are used as the median.

Manly and Francis (2002) used the FPP test based on $\widetilde{Z}_{ij}$ in their two-stage test procedure. In order to carry out the FPP permutation test, they permuted the $\widetilde{Z}_{ij}$ rather than the original observations. This test was quite robust (Francis and Manly, 2001); a bootstrap validation is not required. However, as shown in Section 2.4, rank tests can be more powerful than the FPP test. Thus, it may be useful to apply the WMW (Wilcoxon-Mann-Whitney) test after the Levene transformation, as proposed by Talwar and Gentle (1977) for symmetric distributions.

If the distributions F and G are not symmetric, the WMW test based on the transformed data can have an unacceptably inflated type I error rate, both when performed asymptotically or based on permutations of the transformed values (Neuhäuser, 2004, 2007). In simulations, actual sizes almost as large as 2α were found (Neuhäuser, 2004). For small samples this inflation may be hidden due to the conservatism of the test. Nevertheless, the WMW test after a Levene transformation should not be used when the distribution of the data is unknown or possibly skewed.

For the permutation tests discussed so far in this section, the absolute deviations from sample medians, that is, the transformed values $\widetilde{Z}_{ij}$, are re-

TABLE 3.7: The simulated size of different permutation tests based on permutations of the original observations in case of a pure difference in location

Test	Uniform	Normal	χ^2 (3 df)	Exponential
FPP	≤ 0.001	≤ 0.001	≤ 0.001	≤ 0.001
WMW	0.272	0.302	0.342	0.373

TABLE 3.8: The simulated size of the FPP test based on permutations of the transformed values $\widetilde{Z}_{ij}$

Sample Sizes	Uniform	Normal	χ^2 (3 df)	Exponential
$n_1 = n_2 = 9$	0.015	0.024	0.041	0.047
$n_1 = n_2 = 19$	0.026	0.039	0.048	0.055
$n_1 = 15, n_2 = 9$	0.023	0.029	0.046	0.055

allocated to the samples. However, one can also perform the tests based on permutations of the original observations (Manly 2007, p. 122). Then, a permutation test is based on permutations of the values X_i and Y_j. For each permutation, the Levene transformation is applied separately before the test statistic is calculated. Levene's transformation is carried out after the data are reallocated; the Levene transformation is only one step in the computation of the test statistic. As a permutation test, such a test controls the nominal significance level when there is no difference whatsoever between the groups. Moreover, simulations demonstrated that the FPP and WMW tests are less conservative in this case (Neuhäuser, 2007).

However, the behaviour of such a test based on permutations of the original observations is important in the case of a pure difference in location. Unfortunately, in that case the rank tests cannot control the type I error rate, even for symmetric distributions. Table 3.7 shows this finding for the WMW test. Analogous results can be observed for other linear rank statistics and also for the BWS statistic. In contrast, the FPP test is now extremely conservative, and therefore not very powerful (Neuhäuser, 2007).

As a consequence, the test for a difference in variability should be performed, as suggested by Manly and Francis (2002), as an FPP test based on permutations of the values $\widetilde{Z}_{ij}$. This test has an acceptable size (see Table 3.8). Due to the Levene transformation, size and power do not depend on a possible shift in location. The power of this test is often higher than that of the t test based on $\widetilde{Z}_{ij}$ (Neuhäuser, 2007).

When the sample sizes are unbalanced, that is $n_1 \neq n_2$, the Levene test only asymptotically tests the null hypotheses that the variances are equal (if they exist). Let us consider normally distributed data with the standard deviations σ_1 and σ_2. In this case we have

$$E(Z_{ij}) = \sqrt{\frac{2}{\pi}\left(1 - \frac{1}{n_i}\right)}\ \sigma_i\,,\ i = 1, 2.$$

Hence, the null hypothesis $\frac{n_1-1}{n_1}\sigma_1^2 = \frac{n_2-1}{n_2}\sigma_2^2$ is tested (Keyes and Levy, 1997; O'Neill and Mathews, 2000), which only asymptotically equals $\sigma_1 = \sigma_2$. A corresponding expression holds for the median-based test; the expected values of $\widetilde{Z}_{ij}$, $i = 1, 2$, are multiples of σ_i. The corresponding factor depends on the sample size and converges to $\sqrt{2/\pi}$, for $n_i \to \infty$ (O'Neill and Mathews, 2000).

Keyes and Levy (1997) as well as O'Neill and Mathews (2000) introduced additional transformations for unbalanced data. However, simulation results of Francis and Manly (2001) and Neuhäuser (2007) indicated that such additional transformations are not necessary when a permutation test is performed. The FPP test based on $\widetilde{Z}_{ij}$ was α-robust even for small and unbalanced samples (see also the last row of Table 3.8).

When n_1 and n_2 are both odd, there is at least one $\widetilde{Z}_{ij}$ equal to 0 in each group. For parametric tests and more than two groups, Hines and O'Hara Hines (2000) showed that the deletion of those "structural" zeros prior to performing the statistical test can increase the power. The FPP test, however, can be anticonservative when structural zeros are deleted. According to simulations, the actual size can be larger than 7% in case of $\alpha = 5\%$ (Neuhäuser, 2007). Without presenting details Hines and O'Hara Hines (2000, p. 454) reported an analogous finding: "The observed levels of significance ... proved slightly liberal but were rarely above 10%." A violation of the significance level of this magnitude may be regarded as inacceptable rather than "slightly liberal."

Finally it should be noted that there are many further tests for a difference in variability. Duran (1976) presented a review. However, these tests often violate the significance level if the distributions are very skew and the central locations are different and unknown (Shoemaker, 1995). Here in the nonparametric model, distinctly skew distributions are possible. Furthermore, unknown and therefore potentially unequal central locations are common in applications. The FPP test based on $\widetilde{Z}_{ij}$ can be applied in the case of very skew distributions (Francis and Manly, 2001).

When a location-scale test is performed, the problem mentioned in the previous paragraph is irrelevant. When the locations of the two distributions are unequal, the alternative H_1^{LS} is true. Thus, the Ansari-Bradley test can be applied within a location-scale test although it may be unsuitable as a single test in some situations.

3.4.1   Example and Implementation in SAS

The FPP test based on permutations of the values $\widetilde{Z}_{ij}$ will be applied to the bumblebee example (Table 3.2). The medians of the two groups are $\widetilde{X} = 26$ and $\widetilde{Y} = 32.5$. With the FPP test, a (two-sided) p-value of $2,214/50,388 = 0.0439$ results. A t test carried out with the $\widetilde{Z}_{ij}$ values is not significant at $\alpha = 0.05$.

In order to perform the FPP test based on permutations of the $\widetilde{Z}_{ij}$ in SAS, the procedure NPAR1WAY can be used, as described in Section 2.1. However, the values of $\widetilde{Z}_{ij}$ must be calculated. The medians can be computed with the SAS-procedure MEANS. A sorting with PROC SORT is required in order to determine the medians separately for the two groups:

```
PROC SORT DATA=example1;
  BY group;
RUN;

PROC MEANS MEDIAN;
  VAR count;
  BY group;
  OUTPUT OUT=medians MEDIAN=med;
RUN;

DATA all;
  MERGE example1 medians;
  BY group;
  z_ij=ABS(count-med);
RUN;

PROC NPAR1WAY DATA=all SCORES=DATA;
  CLASS group;
  VAR z_ij;
  EXACT;
RUN;
```

3.5   Concluding Remarks

In particular in the randomization model, it may be appropriate to test with a location-scale test for a difference in location and/or scale. The Lepage test is the nonparametric location-scale test for the two-sample problem commonly applied in statistical practice. In addition to Lepage's test, the Cucconi test was presented in detail. This test, more than 40 years old, but almost unknown in practice, was recently discussed by Marozzi (2008).

Within a closed testing procedure further tests can be performed after a significant location-scale test. To be precise, in a second step, separate tests for a difference in location and a difference in variability can follow (Neuhäuser and Hothorn, 2000). For the test for a shift in location, the tests discussed in Chapter 2 cannot be applied because they do not guarantee the nominal level in the case of potentially unequal variabilities. As possible tests for the nonparametric Behrens-Fisher problem, that is, for testing for a difference in location without assuming equal variances, a bootstrap t test—based on the Welch t statistic—as well as tests proposed by Brunner and Munzel (2000), and Manly and Francis (2002), are discussed. The Brunner-Munzel test can be performed as a permutation test.

The discussed tests for a difference in variability are based on the Levene transformation with sample medians. After this transformation, the Fisher-Pitman permutation test can be applied as suggested by Manly and Francis (2002).

All three tests of this closed testing procedure can be performed with the full nominal level α, an adjustment of the significance level is not necessary (see appendix).

Chapter 4

Tests for the General Alternative

In this chapter we consider the null hypothesis H_0: $F(t) = G(t)$ for all t. Thus, the underlying distributions of the two groups are equal under H_0. However, in contrast to Chapter 2, there are no further restrictions regarding F and G, and H_1: $F(t) \neq G(t)$ for at least one t results as alternative hypothesis. Hence, it is tested whether there is any difference between the two distributions. The test problem can also be formulated one-sided (see below).

4.1 The (Kolmogorov–)Smirnov Test

The standard test for this most general alternative is the Smirnov test, a so-called omnibus test. The Smirnov test is often called the Kolmogorov-Smirnov test, as in the SAS output given below. Because the name Kolmogorov-Smirnov test is so common, it was denoted the (Kolmogorov-)Smirnov test in previous chapters. However, the Kolmogorov-Smirnov test is a goodness-of-fit test that compares one sample with a given distribution. In contrast, the Smirnov test investigates whether two samples come from identical distributions (Berger and Zhou, 2005).

The test statistic of the Smirnov test is the maximum of the absolute value of the difference between the two empirical distribution functions (Büning and Trenkler, 1994, p. 120):

$$T_S = \max_t \left| \hat{F}(t) - \hat{G}(t) \right| .$$

For a sample with the values z_1, z_2, ..., z_n, the empirical distribution function can be calculated as follows:

$$\hat{F}(x) = \begin{cases} 0 & \text{for} \quad x < z_{(1)} \\ i/n & \text{for} \quad z_{(i)} \leq x < z_{(i+1)} \\ 1 & \text{for} \quad x \geq z_{(n)}, \end{cases}$$

where $z_{(1)} < z_{(2)} < \cdots < z_{(n)}$ are the ordered values of the sample. The empirical distribution function can also be determined in the presence of ties. In this case, the function jumps by a multiple of $1/n$, depending on the size of the tied group. Although we do not use ranks to define the test statistic, it should be noted that the Smirnov test is a rank test as the test statistic depends only on the ranks of the observations (Lehmann, 2006, p. 37).

For a one-sided test, T_S is calculated without using the absolute value. Furthermore, if the alternative is $F(t) < G(t)$ (for at least one t), the maximum of the difference $\hat{G}(t) - \hat{F}(t)$ should be used (Büning and Trenkler, 1994, p. 120).

An asymptotic distribution of T_S is tabulated, for instance, by Lehmann (2006, p. 421). However, the approximation with an asymptotic distributions is often poor. On the one hand, the convergence is very slow when the sample sizes are unequal. On the other hand, the asymptotic distribution no longer applies in the presence of ties (Lehmann, 2006, p. 39). Hence, it is recommended to carry out the Smirnov test as a permutation test (Berger and Zhou, 2005). Because the null hypothesis H_0: $F = G$ is investigated, a permutation test in the usual way is possible. Thus, the Smirnov test can be performed as an exact permutation test. In case of large sample sizes, the permutation null distribution can be determined based on a simple random sample of permutations (see Chapter 2).

4.1.1   Example and Implementation in SAS

Wang et al. (1997) compared the genetic variability in the sacred baboon *(Papio hamadryas)* between wild and zoo populations. The mean number of alleles per locus was 1.03 and 1.10, respectively, in the zoos of Frankfurt and Cologne. For five wild populations, the corresponding values were 1.21, 1.21, 1.24, 1.26, and 1.35 (Shoetake, 1981; Wang et al., 1997, p. 156).

First, we need the empirical distribution functions. They are

$$
\hat{F}(x) = \begin{cases} 0 & \text{for} \quad x < 1.03 \\ 0.5 & \text{for} \quad 1.03 \le x < 1.10 \\ 1 & \text{for} \quad x \ge 1.10 \end{cases}
$$

for the baboons living in zoos and

$$
\hat{G}(x) = \begin{cases} 0 & \text{for} \quad x < 1.21 \\ 0.4 & \text{for} \quad 1.21 \le x < 1.24 \\ 0.6 & \text{for} \quad 1.24 \le x < 1.26 \\ 0.8 & \text{for} \quad 1.26 \le x < 1.35 \\ 1 & \text{for} \quad x \ge 1.35 \end{cases}
$$

for the wild populations. In this example, the test statistic T_S is equal to its largest possible value 1. As displayed in Figure 4.1, the empirical distribution function $\hat{F}(x)$ is 1, whereas $\hat{G}(x)$ is 0, between 1.10 and 1.21.

The sample sizes are very small: $n_1 = 2$ and $n_2 = 5$. For the permutation test, $\binom{7}{2} = 21$ permutations must be considered. Within these permutations there is one further permutation with $T_S = 1$. To be precise, this is the permutation for which the five smallest values are in group 2 and the two

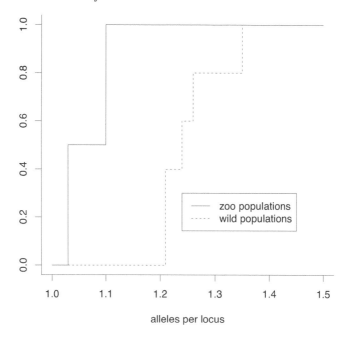

FIGURE 4.1: The empirical distribution functions $\hat{F}(x)$ for the group of baboons living in zoos and $\hat{G}(x)$ for the group of wild baboons.

largest values, 1.26 and 1.35, are in group 1. Consequently, the p-value of the exact Smirnov test is $2/21 = 0.0952$.

The Smirnov test can be performed with the SAS-procedure NPAR1WAY. Within the PROC NPAR1WAY statement, the option KS, or EDF (for "empirical distribution function"), is necessary. For the exact test, the EXACT statement (see Chapter 2) is also needed. Thus, the program is

```
PROC NPAR1WAY EDF;
  CLASS group;
  VAR count;
  EXACT;
RUN;
```

The output of this program presents the two-sided as well as both one-sided Smirnov tests:

```
    Kolmogorov-Smirnov Test for Variable count
          Classified by Variable group


                   EDF at    Deviation from Mean
group        N     Maximum       at Maximum
```

```
1                2        1.000000           1.010153
2                5        0.000000          -0.638877
Total            7        0.285714
```

```
          Maximum Deviation Occurred at Observation 2
               Value of count at Maximum = 1.10
```

```
KS   0.4518    KSa   1.1952
```

```
Kolmogorov-Smirnov Two-Sample Test
```

```
D = max |F1 - F2|      1.0000
Asymptotic Pr >  D     0.1148
Exact       Pr >= D    0.0952
```

```
D+ = max (F1 - F2)     1.0000
Asymptotic Pr >  D+    0.0574
Exact       Pr >= D+   0.0476
```

```
D- = max (F2 - F1)     0.0000
Asymptotic Pr >  D-    1.0000
Exact       Pr >= D-   1.0000
```

Despite the extremely small sample sizes, the asymptotic results are given in the SAS output; the asymtotic two-sided p-value is 0.1148. In case of larger sample sizes, the approximate permutation test with, for example, 10,000 randomly selected permutations can be specified using the option MC (see chapter 2).

4.2   Further Tests

In the preceding example an extreme permutation is observed. In the other possible extreme permutation, the two largest values, 1.26 and 1.35, are assigned to group 1, thus the five smallest values come into group 2. Figure 4.2 shows the empirical distribution functions for this situation.

The Smirnov test looks at the maximum difference T_S between the two empirical distribution functions. This test statistic is also 1 for the extreme permutation displayed in Figure 4.2. However, in this case, the difference between the two empirical distribution functions is 1 only between 1.24 and 1.26. For the observed data, the difference is 1 for the much larger range from 1.10

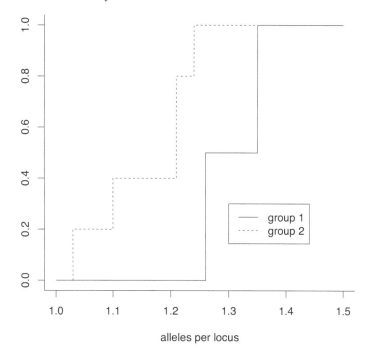

FIGURE 4.2: The empirical distribution functions for the extreme permutation where the two largest values, 1.26 and 1.35, are in group 1 and the five smallest values, 1.03, 1.10, 1.21, 1.21, and 1.24, are in group 2.

to 1.21 (see Figure 4.1). The test statistic T_S cannot see any difference between these situations. Other tests can. Schröer and Trenkler (1995) proposed the alternative approach to consider the area between the two empirical distribution functions as the test statistic. They did not present any distribution of their test statistic; however, the test can be performed as a permutation test.

The BWS and Z_C tests introduced in Chapter 2 were also proposed for the general alternative $F \neq G$. They could be investigated in Chapter 2 for the location problem because it was assumed that the distributions F and G are identical with only the exception of a possible location shift. Simulation results of Baumgartner et al. (1998) and Zhang (2006) demonstrate that the tests BWS and Z_C may be more powerful than the Smirnov test.

When applying the BWS test to the example data about the genetic variability in the sacred baboon, the test statistic B has its maximum at 3.3388 for the observed data. The statistic B is smaller for all other possible permutations. Hence, the exact BWS test has the p-value $P_0(B \geq 3.3388) = 1/21 = 0.0476$, which is significant for $\alpha = 5\%$. For the other extreme permutation, that is, when the two largest values are assigned to group 1, we have $B = 2.3531$.

The Z_C test is also significant at $\alpha = 5\%$. We have $Z_C = 1.2861$ for the observed data. This is the minimum of Z_C, and no other permutation has such a small value of the test statistic. Thus, the p-value is $P_0(Z_C \leq 1.2861) = 1/21 = 0.0476$. Please note that small values of Z_C provide evidence against the null hypothesis. For the other extreme permutation (the two largest values are in group 1), we have $Z_C = 1.3107$.

Chapter 5

Ordered Categorical and Discrete Numerical Data

Ordinal data are common in practice. The values of ordinal or ordered categorical variables can be arranged according to size. However, calculations such as addition and subtraction are not possible (see appendix). An ordinal random variable usually has only a few mass points. The three examples with ordinal data presented by Brunner and Munzel (2002) have four, five, and ten possible values, respectively. The sample sizes are often much larger; hence ties are not only possible, but inevitable.

The situation is similar for discrete numerical variables. Because such data are numerical, differences can be computed. Examples are counts such as the number of children. Again, there are often a few possible values only. Table 5.1 shows an example: a case-control study that investigates a possible association between an allele (gene) and a specific disease. The cases are a random sample of persons with the disease, whereas the control group is a random sample of persons without the disease. For each person, the number of alleles A is gathered. This allele A is suspected of causing or promoting the disease. For diploid organisms, only three results are possible: 0, 1, or 2 copies of the allele A. Hence, there are a lot of ties. A real data set is presented with an SAS program in Section 7.5.

5.1 Tests for Comparing Multinomial Distributions

The comparison of two groups with ordinal or discrete numerical data is a comparison of two multinomial distributions. The Wilcoxon-Mann-Whitney (WMW) test is commonly suggested for such a comparison (e.g., Rahlfs and Zimmermann, 1993; Nanna and Sawilowsky, 1998). A linear rank test such as the WMW test, however, can have a very low power for parts of the alternative (Berger and Ivanova, 2002).

Due to the many ties, it is almost impossible to assess whether or not an asymptotic distribution is appropriate: "The number of ties in each category, the group imbalance, and the choice of rank scores all affect the shape of the permutation distribution in complicated ways, making it difficult to predict whether the asymptotic results for a given data set are reliable" (Mehta et al., 1992, p. 22). A permutation test should therefore be performed rather than an asymptotic test. Instead of the WMW or other linear rank tests, the exact Baumgartner-Weiss-Schindler (BWS) test can also be applied. The Fisher-Pitman permutation (FPP) test cannot be carried out for ordinal data, as already mentioned in Chapter 2. For discrete numerical data, however, the FPP test can be used. Therefore, this test is included in the following study of tests for the comparison of two multinomial distributions.

Please note that the standard test for the analysis of case-control data such as the 2x3-contingency table presented in Table 5.1 is the Armitage test (Armitage, 1955; Slager and Schaid, 2001), sometimes called the Cochran-Armitage test (see, e.g., Neuhäuser and Hothorn, 1999). We will demonstrate that the Armitage test is equivalent to the FPP test. Using the notation as defined by Table 5.1, the test statistic of the Armitage test is $Z = U/\sqrt{\widehat{\text{Var}}(U)}$ with

$$U = \sum_{i=0}^{2} x_i \left(\frac{n_2}{N} r_i - \frac{n_1}{N} s_i \right) \text{ and}$$

$$\widehat{\text{Var}}(U) = \frac{n_1 n_2}{N^3} \left(N \sum_{i=0}^{2} x_i^2 (r_i + s_i) - \left(\sum_{i=0}^{2} x_i (r_i + s_i) \right)^2 \right),$$

where x_i, $i = 0, 1, 2$, are arbitrary scores. For instance, one can use the number of alleles A as the score, that is, $x_i = i$ (Slager and Schaid, 2001). The Armitage test can be carried out as a permutation test or as an asymptotic test. The standardized statistic Z is, under the null hypothesis, asymptotically standard normal. However, the normal approximation may become unreliable when expected frequencies are small (Williams, 1988; Chen et al., 1997).

Within a conditional test the marginal totals of the contingency table are regarded as fixed. Anyway, each permutation of the observed data results in a contingency table with the same marginal totals as the observed table. The variance estimation is identical for all 2x3-tables with equal marginal totals. Hence, the permutation test can also be performed with the statistic U. With the scores (0, 1, 2) mentioned above we have

$$U = \frac{1}{N} \left(n_2 r_1 - n_1 s_1 + 2 n_2 r_2 - 2 n_1 s_2 \right).$$

The difference in the mean number of the allele A per person between the two groups (cases and controls) is

$$\Delta = \frac{r_1 + 2 r_2}{n_1} - \frac{s_1 + 2 s_2}{n_2} = \frac{1}{n_1 n_2} \left(n_2 r_1 + 2 n_2 r_2 - n_1 s_1 - 2 n_2 s_2 \right).$$

This difference Δ is proportional to the statistic U; U and Δ differ by a constant factor that only depends on the sample sizes. Because Δ can be used

TABLE 5.1: Frequency distribution of the genotypes aa, aA, and AA in a case-control study

	aa	aA	AA	Total
Cases	r_0	r_1	r_2	n_1
Controls	s_0	s_1	s_2	n_2
Total	$r_0 + s_0$	$r_1 + s_1$	$r_2 + s_2$	N

as the test statistic of the FPP test (see Section 2.1), the exact Armitage test in this situation is equivalent to the FPP test (Neuhäuser, 2002d).

In this chapter the null hypothesis H_0: $F(t) = G(t)$ for all t is investigated again. Hence, a permutation test in the usual way is possible. The alternative is H_1: $F(t) \neq G(t)$ for at least one t. If the alternative is true, there may be a location shift. However, a constant shift in the mass points of a discrete distribution is rare in practice (Brunner and Munzel, 2002, p. 49).

In the two-sample case, ordinal or discrete numerical data with k possible values/categories can be presented in a 2xk-contingency table (see Table 5.2). Each permutation results in a 2xk table with the same marginal totals. The vectors of the probabilities of the k categories in groups 1 and 2 are denoted by $\pi_1 = (\pi_{11}, \pi_{12}, \ldots, \pi_{1k})$ and $\pi_2 = (\pi_{21}, \pi_{22}, \ldots, \pi_{2k})$. The elements of each of these two row vectors sum to 1. The null hypothesis $F = G$ can also be expressed in the form $\pi_1 = \pi_2$. The alternative $F \neq G$ is equal to $\pi_1 \neq \pi_2$.

Because the test problem H_0: $F = G$ versus the general alternative H_1: $F \neq G$ is considered, we also investigate Pearson's χ^2 test as well as the Smirnov test (Neuhäuser, 2003c). These two tests can be performed as exact permutation tests. The χ^2 test compares the actually observed cell counts with the counts expected under H_0. However, this test does not utilize the ordinal scale of the result variable.

5.2 The Exact χ^2 Test

The test statistic of the χ^2 test is

$$X^2 = \sum_{i=1}^{2} \sum_{j=1}^{k} \frac{\left(x_{ij} - \frac{n_i m_j}{N}\right)^2}{\frac{n_i m_j}{N}},$$

and the notation is defined by Table 5.2. The test statistic X^2 is, under the null hypothesis, asymptotically χ^2 distributed with $k - 1$ degrees of freedom for a 2xk-contingency table.

TABLE 5.2: Notation for a $2 \times k$ contingency table

	Result 1	Result 2	...	Result k	Total
Group 1	x_{11}	x_{12}	...	x_{1k}	n_1
Group 2	x_{21}	x_{22}	...	x_{2k}	n_2
Total	m_1	m_2	...	m_k	N

TABLE 5.3: Example data for a 2×4 contingency table

	Result 1	Result 2	Result 3	Result 4	Total
Group 1	0	1	4	3	8
Group 2	2	1	1	4	8
Total	2	2	5	7	16

In case of dichotomous data, we have $k = 2$, a 2x2-table, and hence a comparison of two binomial distributions. In that case, the test statistic of the χ^2 test differs from the square of the WMW test statistic by the constant factor $N/(N-1)$ only (Brunner and Munzel, 2002, p. 67). A further test for the comparison of two binomial distributions is Fisher's exact test. As its test statistic, a cell count of the 2x2-table can be used. This statistic can be expressed as a strictly monotonic transformation of the test statistic of the exact WMW test. Thus, the two test statistics are equivalent for a permutation test, and Fisher's exact test is a special case of the exact WMW test (Brunner and Munzel, 2002, p. 65).

Let us consider again the example from the Old Testament (Section 2.7). The four Israelites were better fed than the four servants. Thus, the following 2x2-table results:

$$
\begin{array}{c|c}
4 & 0 \\
\hline
0 & 4
\end{array}
$$

With these data we have $X^2 = 8$ and the exact (two-sided) p-value of the χ^2 test is 0.0286. Fisher's exact test also has the exact two-sided p-value 0.0286 (computed using the method usually proposed, see, e.g., Ruxton and Neuhäuser, 2010b). These p-values are identical to the exact two-sided p-value of the WMW test (see Section 2.7).

Due to the multitude of ties within a 2xk-table, an exact permutation test can be carried out even for large sample sizes. We illustrate this using the example presented in Table 5.3

With the data of Table 5.3 we get $X^2 = 3.943$. An asymptotic χ^2 test would give a p-value of 0.2677. However, the contingency table is too sparse to use the asymptotic distribution; all expected cell counts are less than 5. Thus, the common rule of thumb is violated. According to this rule of thumb, an asymptotic test is possible if all expected counts of a 2x2-table are at least 5; for a larger contingency table at least 80% of the expected counts should be 5 or larger; in addition, no expected counts should be less than 1 (Siegel, 1956, p. 110). Furthermore, the only reason to prefer asymptotic over exact testing is ease of calculation, an advantage that is declining with advances in computing power (Ruxton and Neuhäuser, 2010b). Consequently, a permutation test should be performed for the data of Table 5.3.

Permutations of the observations do not change the marginal totals n_i and m_j. Thus, the marginal totals are constant, and a conditional test results. In the example (Table 5.3), fifty-two different tables are possible in order to distribute the sixteen observations into the eight cells of a 2x4-contingency table with marginal totals as in Table 5.3. An algorithm that can be used to list these 52 tables is presented below in this chapter.

The possible tables with identical marginal totals are not equiprobable. Under H_0, the probability for a table is (Weerahandi, 1995, pp. 99–102),

$$P_0(r_{11}, \ldots, r_{2k}) = \frac{n_1! \, n_2! \displaystyle\prod_{j=1}^{k} m_j!}{N! \displaystyle\prod_{j=1}^{k} r_{1j}! \, r_{2j}!},$$

where the r_{ij} denote the cell counts of the generated table. For the fifty-two possible tables, these probabilities range between 0.00016 and 0.10878.

Why is there such a big difference between the probabilities of the tables? If there are no ties, then there are $\binom{16}{8} = 12,870$ different permutations in the case of $n_1 = n_2 = 8$. However, here there are a lot of ties because the variable "result" can take on four different values only. Hence, not all of the 12,870 permutations are different.

Let us consider the two tables with the extreme probabilities. The table with the smallest probability 0.00016 is:

	Result 1	Result 2	Result 3	Result 4
Group 1	1	2	5	0
Group 2	1	0	0	7

In three of four columns, all observations are in one row. Thus, no other allocation of those observations is possible without changing the cell counts. However, in the first column there is one observation in each of the two rows. Here, these two observations could be allocated the other way round, too— without changing the contingency table. Hence, there are $\binom{1+1}{1} = 2$ permutations that give the above table. Therefore, the probability of this table is $2/12,870 = 0.00016$.

Let us now consider the table with the largest probability 0.10878:

	Result 1	Result 2	Result 3	Result 4
Group 1	1	1	2	4
Group 2	1	1	3	3

In this table there is no cell with the count 0. Hence, there are many more permutations that lead to this table. To be precise, there are, in total, $\binom{1+1}{1}\binom{1+1}{1}\binom{2+3}{2}\binom{4+3}{4} = 1,400$ permutations. Thus, the probability for this table is $1,400/12,870 = 0.10878$.

The p-value of the exact test is the sum of the probabilities of those tables that give at least as much evidence against the null hypothesis as the actually observed table. In the example, the p-value of the χ^2 permutation test therefore is the sum of the probabilities of the tables for which we have $X^2 \geq 3.943$. This sum is $P_0(X^2 \geq 3.943) = 0.3691$.

The exact χ^2 test is a usual permutation test: The test statistic is Pearson's χ^2 statistic. Because there are a lot of ties, many permutations result in the same contingency tables. Therefore, it is possible to consider the relatively small number of different contingency tables (with constant marginal totals) rather than the huge number of possible permutations. In the example, one can compute the exact p-value based on 52 tables rather than based on 12,870 permutations. That is the reason why an exact test is still possible for relatively large sample sizes. However, one must take into consideration that the different tables have different probabilities under the null hypothesis. These different probabilities are caused by the fact that the number of permutations that lead to a specific table differs.

The possible contingency tables, that is, all possible tables with fixed marginal totals, can be generated using the following algorithm according to Williams (1988):

The observed cell counts are denoted by x_{ij} ($i = 1, 2$; $j = 1, 2, \ldots, k$); see Table 5.2. The cell counts of a generated table are analogously denoted by r_{ij}. We have $\sum_{i=1}^{k} x_{1i} = n_1$ and $\sum_{i=1}^{k} x_{2i} = n_2$. Now, the cell count r_{11} can range from

$$\max\left(0, n_1 - \sum_{i=2}^{k} m_i\right) \text{ to } \min(m_1, n_1),$$

r_{21} can be calculated as $r_{21} = m_1 - r_{11}$. With the given values of r_{11} and r_{21}, the range of r_{12} is

$$\max\left(0, n_1 - r_{11} - \sum_{i=3}^{k} m_i\right) \text{ to } \min(m_2, n_1 - r_{11}),$$

and r_{22} is $r_{22} = m_2 - r_{12}$. With the given values of $r_{11}, \ldots, r_{1\,j-1}$, $r_{21}, \ldots, r_{2\,j-1}$, the range of r_{1j} ($j = 3, \ldots, k - 1$) is

$$\max\left(0, n_1 - \sum_{i=1}^{j-1} r_{1i} - \sum_{i=j+1}^{k} m_i\right) \text{ to } \min\left(m_i, n_1 - \sum_{i=1}^{j-1} r_{1i}\right),$$

and $r_{2j} = m_j - r_{1j}$. In the last step, we calculate $r_{1k} = n_1 - \sum_{i=1}^{k-1} r_{1i}$ and

$r_{2k} = m_k - r_{1k}$. For each table the test statistic must be computed. The p-value of the exact test is then the probability for those tables that give at least as much evidence against the null hypothesis as the observed table.

5.2.1 Implementation in SAS

In SAS, Pearson's χ^2 test can be carried out with the procedure FREQ as follows:

```
DATA example53;
  INPUT group result count;
  CARDS;
  1 1 0
  1 2 1
  1 3 4
  1 4 3
  2 1 2
  2 2 1
  2 3 1
  2 4 4
  ;

PROC FREQ;
  TABLES group*result;
  WEIGHT count;
  EXACT CHISQ;
RUN;
```

The statement EXACT CHISQ is necessary in order to perform an exact permutation test. An approximate permutation test is possible with EXACT CHISQ / MC. When only an asymptotic χ^2 test is required, the EXACT statement can be omitted. Instead, one must specify CHISQ within the TABLES statement:

```
TABLES group*result / CHISQ;
```

In order to apply the WMW test or another test using the procedure NPAR1WAY, the variable that includes the cell counts of the contingency table can be specified in the FREQ statement:

```
PROC NPAR1WAY WILCOXON;
  CLASS group;
  VAR result;
  EXACT;
  FREQ count;
RUN;
```

In case a test statistic not implemented in a procedure should be applied, one must generate all possible contingency tables, that is, all tables with marginal totals identical to the observed table. This can be done using the algorithm of Williams (1988). The probabilities of the different tables can be computed very fast by means of a logarithmic transformation; see Williams (1988) for details.

5.3   Comparison of the Tests

The size of five different tests is displayed in Table 5.4; in addition to the FPP, WMW, and BWS tests, the Smirnov and the χ^2 test are investigated. Because there are only k categories, may ties occur, however, the pattern of ties is random and, hence, the size is simulated even for the rank tests. In order to apply the FPP test, numerical values are required: We use 1, 2, ..., k for the simulation. Of course, this is arbitrary; other values are also possible. For the case-control data of Table 5.1 the first category has the value 0 (number of alleles A). In Section 7.4 an example is presented where the values are 3, 4, and

TABLE 5.4: The simulated size for different permutation tests

Number of Possible Categories	Probabilities $\pi_{11}, \pi_{12}, \ldots, \pi_{1k}$[a]	FPP	WMW	BWS	KS[b]	χ^2
Sample sizes: $n_1 = n_2 = 10$						
$k = 3$	$\frac{1}{2}, \frac{1}{4}, \frac{1}{4}$	0.035	0.030	0.033	0.017	0.037
$k = 6$	$\frac{1}{2}, \frac{1}{10}, \frac{1}{10}, \frac{1}{10}, \frac{1}{10}, \frac{1}{10}$	0.040	0.044	0.046	0.026	0.041
$k = 3$	$\frac{1}{3}, \frac{1}{3}, \frac{1}{3}$	0.034	0.033	0.036	0.019	0.038
$k = 4$	$\frac{1}{4}, \frac{1}{4}, \frac{1}{4}, \frac{1}{4}$	0.036	0.038	0.041	0.018	0.047
$k = 5$	$\frac{1}{5}, \frac{1}{5}, \frac{1}{5}, \frac{1}{5}, \frac{1}{5}$	0.038	0.043	0.044	0.022	0.045
$k = 6$	$\frac{1}{6}, \frac{1}{6}, \frac{1}{6}, \frac{1}{6}, \frac{1}{6}, \frac{1}{6}$	0.038	0.044	0.046	0.022	0.043
Sample sizes: $n_1 = 5, n_2 = 10$						
$k = 5$	$\frac{1}{5}, \frac{1}{5}, \frac{1}{5}, \frac{1}{5}, \frac{1}{5}$	0.038	0.036	0.047	0.028	0.039
Sample sizes: $n_1 = 10, n_2 = 20$						
$k = 5$	$\frac{1}{5}, \frac{1}{5}, \frac{1}{5}, \frac{1}{5}, \frac{1}{5}$	0.042	0.045	0.049	0.028	0.047

[a] $\pi_1 = \pi_2$ under H_0.
[b] KS = Smirnov test.

TABLE 5.5: The simulated power of different permutation tests for location-shift alternatives

Probabilities $\pi_{i1}, \pi_{i2}, \ldots, \pi_{ik}$	n_1	n_2	FPP	WMW	BWS	KS[a]	χ^2
Number of possible categories: $k = 6$							
Group 1: 0.6, 0.1, 0.1, 0.1, 0.1, 0							
Group 2: 0, 0.6, 0.1, 0.1, 0.1, 0.1	5	10	0.23	0.41	0.72	0.49	0.75
	10	10	0.28	0.53	0.79	0.67	0.90
	10	20	0.40	0.64	0.99	0.84	0.99
Group 1: $\frac{1}{4}, \frac{1}{4}, \frac{1}{4}, \frac{1}{4}, 0, 0$							
Group 2: $0, 0, \frac{1}{4}, \frac{1}{4}, \frac{1}{4}, \frac{1}{4}$	5	10	0.81	0.75	0.82	0.49	0.42
	10	10	0.97	0.94	0.96	0.73	0.73
Number of possible categories: $k = 5$							
Group 1: 0.7, 0.1, 0.1, 0.1, 0							
Group 2: 0, 0.7, 0.1, 0.1, 0.1	5	10	0.36	0.60	0.88	0.75	0.87
	10	10	0.49	0.74	0.94	0.89	0.97

[a]KS = Smirnov test.

5. However, only the FPP test depends on the exact values. The mentioned equidistant values, such as $(1, 2, 3)$ or $(3, 4, 5)$, give identical p-values. Other, not equidistant choices can change the FPP p-value. However, the differences are, at least for the t test, often small (Labovitz, 1970).

The actual level of the BWS test is closer to the nominal α than that of the FPP, WMW, and Smirnow tests. The FPP test has a larger size for one simulated scenario. Note that the conservatism of the Smirnov test in the case of ties is well known (Büning and Trenkler, 1994, p. 122). The size of the χ^2 test is sometimes smaller and sometimes larger than that of the BWS test.

Tables 5.5 and 5.6 show power results. The BWS test is more powerful than the FPP, WMW, and Smirnov tests both for location shifts (Table 5.5) as well as for general alternatives (Table 5.6). The difference in power is often large. A comparison between the BWS and the χ^2 tests does not show a clear winner. The BWS test, however, has the advantage that it can be applied irrespective whether and how many ties occur. In contrast, when there are few ties, the χ^2 test depends on the number and the definition of the formed categories. Obviously, this "binning" of data is arbitrary, and not necessary for the BWS test.

In the case of general alternatives, it is possible that the distribution functions F and G cross, as in case of $k = 3$ in Table 5.6. Horn (1990) demonstrated that the WMW test can have a very low power in such a situation. The power of the BWS and the χ^2 tests, however, is relatively large even in this case.

TABLE 5.6: The simulated power of different permutation tests for general alternatives ($\alpha = 0.05$)

Probabilities $\pi_{i1}, \pi_{i2}, \ldots, \pi_{ik}$	n_1	n_2	FPP	WMW	BWS	KS[a]	χ^2
Number of possible categories: $k = 3$							
Group 1: 0.6, 0.2, 0.2							
Group 2: 0.1, 0.8, 0.1	5	10	0.22	0.35	0.56	0.41	0.51
	10	10	0.21	0.34	0.55	0.46	0.70
	10	20	0.38	0.51	0.88	0.80	0.87
Number of possible categories: $k = 6$							
Gr. 1: 0.5, 0.1, 0.1, 0.1, 0.1, 0.1							
Gr. 2: 0.1, 0.5, 0.1, 0.1, 0.1, 0.1	10	10	0.07	0.16	0.30	0.23	0.50
Gr. 2: 0.1, 0.1, 0.5, 0.1, 0.1, 0.1	10	10	0.17	0.24	0.38	0.31	0.50
Gr. 2: 0.1, 0.1, 0.1, 0.5, 0.1, 0.1	10	10	0.33	0.36	0.46	0.38	0.50
	5	10	0.27	0.25	0.38	0.30	0.35
	10	20	0.48	0.47	0.64	0.57	0.63
Gr. 2: 0.1, 0.1, 0.1, 0.1, 0.5, 0.1	10	10	0.48	0.48	0.55	0.43	0.50
Gr. 2: 0.1, 0.1, 0.1, 0.1, 0.1, 0.5	10	10	0.58	0.63	0.63	0.47	0.50

[a]KS = Smirnov test.

For a one-sided test problem, the alternative can be defined as follows (Berger and Ivanova, 2002):

$$H_1^> : \sum_{i=1}^{l} \pi_{1i} \geq \sum_{i=1}^{l} \pi_{2i} \text{ for all } l = 1, \ldots, k \text{ and } \pi_1 \neq \pi_2 .$$

Under $H_1^>$ crossing distribution functions are impossible. Berger and Ivanova (2002) showed that, in general, no optimal test exists for H_0 versus $H_1^>$. Nevertheless, the one-sided BWS test based on B^* (see Section 2.5) has a relatively large power. The results for the one-sided test problem are not presented here, but they are similar to those presented in Tables 5.4 through 5.6 (Neuhäuser, 2005b).

In summary, exact tests based on B and B^*, respectively, can also be suggested for ordinal and discrete numerical data (Neuhäuser, 2003c). The BWS test is less conservative and more powerful than the often applied WMW test. Moreover, according to simulation results, the BWS test is more powerful than the Smirnov and FPP tests. The latter test cannot be used for ordinal data. The comparison between the BWS and χ^2 tests did not show a clear winner, but the BWS test has the advantage that it is not necessary to form categories. For the χ^2 test, however, an arbitrary binning of data might be necessary.

Chapter 6

The Conservatism of Permutation Tests

> "Choosing a conservative test is equivalent to discarding data which may have been collected at considerable cost."
> (Williams, 1988, p. 431).

As discussed in Chapter 2, the randomization model is often more appropriate than the population model, as it might be for a randomized clinical trial, which is usually based on a "convenience sample" rather than a random sample. In that case, a permutation test is the "platinum standard" (Tukey, 1993). However, in practice there are situations where a permutation test is not performed although it is doable and appropriate. Berger (2009) discussed and criticized this in a Socratic dialogue where Socrates asked: "If you can observe the exact p-value, then why would you go on to attempt to approximate it?"

6.1 Disadvantages of Bootstrap and Permutation Tests

Permutation tests also have disadvantages. On the one hand, permutation tests are computer intensive, as there are a huge number of possible permutations in the case of large samples. Although this point is more pronounced in the case of more than two groups (see Chapter 9), it is also relevant for the two-sample problem. For instance, for $n_1 = n_2 = 20$, there are more than 137 billion permutations (1 billion is defined here as 10^9). Obviously, bootstrap methods are computer intensive too. However, the disadvantage is declining over time. Very efficient algorithms were developed (see, e.g., Good, 2000, Chapter 13). Moreover, advances in computer power are huge. Modern PCs probably were unimaginable for R. A. Fisher when he invented permutation tests in the 1930s. In addition, there is the possibility of performing approximate permutation tests based on a random sample of permutations.

The second disadvantage of permutation tests is their conservatism. The reason is the discreteness of the permutation null distribution, especially in the case of small sample sizes. "This conservatism, which is entirely attributable to the discreteness of the test statistic, is the price you pay for exactness" (Cytel, 2007, p. 1231). Ryman and Jorde (2001, p. 2371) wrote that permutation tests are "necessarily conservative"; nevertheless, they suggest applying them whenever possible. The conservatism is the reason that "the virtues of permutation tests continue to be debated in the literature" (Berger, 2000, p. 1319). Of course, it is the "ensuing loss of power, and not the conservatism itself, that is a concern" (Berger, 2000, p. 1325).

6.2   Approaches to Reduce the Conservatism

During the past decades, several approaches were proposed for how conservatism can be reduced. One possibility is a randomized test. In randomized tests, an additional auxiliary Bernoulli trial might decide whether or not the null hypothesis is rejected (for details see, e.g., Mood et al., 1974, p. 404). This principle, however, is not acceptable for statistical practice (Mehta and Hilton, 1993; Senn, 2007) as the decision does not only depend on the observations and therefore is not always reproducible. At first glance one may argue that the same applies to approximate permutation tests and almost all bootstrap methods. They depend on a random sample of permutations or bootstrap samples. Are these tests reproducible? Yes they are, provided that the sample of permutations, or bootstrap samples, is not too small. One can calculate confidence intervals for the p-value; and when the resulting p-value is close to α, the random sample of permutations can be enlarged. This enlargement can formally be performed as a sequential test (Lock, 1991).

For a randomized test, this is not possible. When for some value of the test statistic the success probability of the auxiliary Bernoulli trial is not too far away from 0.5, the result is not reproducible.

Another relatively old proposal is a so-called mid-p-value (Lancaster, 1961). The usual p-value is the probability $P_0(T \geq t^*)$, where T is the test statistic and t^* its observed value, and P_0 denotes the probability under the null hypothesis. A mid-p-value is determined by subtracting half of the probability for the observed value from the exact p-value. To be precise, the mid-p-value is defined as $P_0(T > t^*) + 0.5P_0(T = t^*)$ (see, e.g., Agresti, 2003). Such a mid-p-value is exactly the center of the p-value-interval $[P_0(T > t^*), P_0(T \geq t^*)]$ defined by Berger (2000, 2001). When using the mid-p-value, the decision only depends on the observations, in contrast to a randomized test. However, the disadvantage is that the nominal significance level cannot be guaranteed when a mid-p-value is used (see, e.g., Agresti, 2003). Thus, the approach cannot be suggested in general.

When performing a two-sample comparison, the observed data can be displayed in a 2xk contingency table. When there are no ties, the table has the dimension 2xN. In this case, all cell counts are 0 or 1, and the marginal totals for all columns are 1, that is, $m_i = 1$ for all i (using the notation defined in Table 5.2). In the presence of ties, the number of columns is reduced: $k < N$, and $m_i > 1$ for at least one column. As a consequence, different tables have different probabilities (see Chapter 5). Now one can exclude those tables from the calculation of the p-value that give the same value of the test statistic, but have a larger probability than the observed table (Chen et al., 1997). "Exclude" means here that these tables are regarded as giving less evidence against the null hypothesis; hence, the probabilities of these tables are not considered for the summation when computing the p-value. This approach is acceptable with regard to compliance with the nominal level (Agresti, 2003) because we have $P_0(p\text{-value} \leq \alpha) \leq \alpha$ for $0 < \alpha < 1$.

Nevertheless, this approach is rarely applied in practice. The main reason might be the justification for excluding tables with the same value of the test statistic as observed. Actually, these tables give as much evidence against the null hypothesis as the observed table; therefore, they should be considered for the p-value computation. However, it should be noted that the usually recommended way to perform Fisher's exact test in a two-sided way is based on the probabilities of the tables. To be precise, all tables with probabilities less than or equal to that of the observed table are considered for summing up the p-value.

In the latter approach suggested by Chen et al. (1997), the probability is a *back-up* statistic that can order permutations within a tied group of the test statistic. As mentioned in Chapter 2, in a permutation test the test statistic orders the permutations. However, permutations with an identical value of the test statistic cannot be ordered. A back-up statistic is a second statistic that is used to order the permutations within tied groups of the first statistic. This approach led to a p-value within the p-value-interval defined by Berger (2000, 2001). Without a back-up statistic, the p-value is equal to the upper bound $P_0(T \geq t^*)$ of the interval. Streitberg and Röhmel (1990) presented an example with the Wilcoxon rank-sum as the first test statistic and the Ansari-Bradley statistic as back-up. Cohen and Sackrowitz (2003) recommended back-up statistics for the analysis of contingency tables.

When an exact permutation test is conservative, the actual level is smaller than the nominal α. Boschloo (1970) suggested the use of a new level, a so-called raised conditional level γ (with $\gamma \geq \alpha$), and to reject the null hypothesis if the usual (conditional) p-value is $\leq \gamma$. This is a valid procedure; however, this is an unconditional test (see Lydersen et al. 2009); that is, not all marginal sums are regarded as fixed. A disadvantage is that the raised conditional level γ depends not only on the used test statistic, but also on α, the sample sizes, and the observed pattern of ties. For Fisher's exact test, Boschloo (1970) presented extensive tables for γ.

For an unconditional test, one can assume no marginal totals as fixed, or only one of the margins (either columns or rows). In the two-sample case, the data can be displayed in a $2 \times k$-contingency table; then the marginal sums of the two rows are fixed; these two sums are the sample sizes. In a conditional test, one considers only tables with identical marginal totals, as observed for generating the permutation null distribution. Thus, the marginal totals of the columns are also regarded as fixed although they are not fixed by the design. This is different in an unconditional test; thus, one also considers tables with other column totals. In the two-sample situation, it means that the occurrence and frequency of ties can vary, as in a bootstrap test. In a conditional test, only tables with an identical pattern of ties are considered; the null distribution is "the conditional distributions of the statistics concerned given that the number of observations in each tied group is a fixed constant" (Putter, 1955, p. 368).

Which approach, conditional or unconditional, is more appropriate? For the comparison of binomial distributions, this topic has been controversially discussed since the 1940s (Mehta and Hilton, 1993). This debate is not closed (see, e.g., Mehrotra et al., 2003; Proschan and Nason, 2009): the "controversy ... is still unresolved because ultimately the choice is philosophical rather than statistical" (Mehta and Hilton, 1993, p. 91).

The main reason for an unconditional test is that the null distribution is often less discrete and, hence, the conservatism is reduced and the power is increased, especially for a 2x2-table. However, supporters of the conditional test, including the well-known statisticians R. A. Fisher and D. R. Cox, argue "that only experimental outcomes of equal information should be included in the reference set of possible outcomes. For this reason it is argued that S [i.e. the number of successes] should be fixed at its observed value in hypothetical repetitions of the experiment" (Mehta and Hilton, 1993, p. 92).

In the randomization model there is an additional argument for the conditional approach. Because there is no random sample, one can regard the experimental units, for example the patients in a clinical trial, as fixed. The allocation of these experimental units to the two groups does not affect the column totals under H_0. Thus, it is justified to condition on the observed column totals (Cox, 1958; Berger, 2000).

As shown in previous chapters, tests based on the statistics B and B^* are hardly conservative, even in the case of small samples. Alternative tests are often more conservative. Thus, one main reason to apply an unconditional test, or to use any other method to reduce the conservatism, is omitted, or at least weakened. Let us consider an example. The sample size is 10 per group, there are no ties, and the ranks of the observations from group 1 are 4, 6, 8, 13, 15, 16, 17, 18, 19, and 20. This example comes from the comparison of groups 2 and 3 in Neuhäuser and Bretz (2001, p. 579). The WMW and BWS tests give the following results: $W = 136$ with p $= 0.0185$ and $B = 3.7582$ with p $= 0.0146$. In order to compute the mid-p-value, half of the probability $P_0(T = t^*)$ is subtracted from the usual p-value. This probability strongly differs between the two tests:

$$P_0(W = 136) = 356/184,756 = 0.0019 \text{ and}$$
$$P_0(B = 3.7582) = 2/184,756 = 0.00001.$$

Many software packages display p-values with four decimal places. In that case, one cannot see a difference between the usual and the mid-p-value as far as the BWS test is applied. For the WMW test, three decimal places are sufficient to see a difference.

Moreover, the potential influence of a back-up statistic is very different. Regarding the BWS test, only the two permutations with $B = 3.7582$ can be sorted by a back-up statistic. Hence, a back-up statistic has a very minimal effect. For the WMW test, however, 356 permutations with $W = 136$ are tied without a back-up statistic.

The extent of conservatism can be quantified with the p-value-interval of Berger (2001), independent of α. The upper bound of this interval is the usual p-value. For the lower bound, the probability that the test statistic is equal to its observed value is subtracted. For the example discussed above, the p-value-interval is $[0.0166, 0.0185]$ for the WMW test. For the BWS test, the p-value-interval is $[0.0146, 0.0146]$. The length of this interval is 0.00001; hence, four decimal places are not sufficient to see that it is indeed an interval. The p-value-interval also shows the much smaller conservatism of the BWS test in comparison with the WMW test.

Chapter 7

Further Examples for the Comparison of Two Groups

In the following, the methods discussed in Chapters 2 to 5 are illustrated using data sets from a variety of applications. First, an educational experiment and a clinical trial are considered. The third example presents a comparison of soil lead values from two different districts, and brood sizes of birds are considered in the fourth example. Moreover, an epidemiological case-control study is analyzed.

7.1 A Difference in Location

Williams and Carnine (1981) compared two different methods of training preschool children. Fourteen children were randomized to two groups: an experimental group and a control group. For each child, the number of correct identifications among eighteen new examples was observed; see Williams and Carnine (1981) for further details. Raw data were given by Gibbons (1993, p. 31) and are presented in Table 7.1.

In this example we do not have random samples from defined populations. Instead, the groups are created by randomization. Hence, the randomization model and permutation tests are appropriate. Furthermore, permutation tests are also useful because of the small sample size, in combination with ties.

The box plots are displayed in Figure 7.1. One can see distinct differences between the groups, especially in location. The location-scale tests give

TABLE 7.1: The number of correct identifications in two groups of preschool children

Experimental group	15	18	8	15	17	16	13
Control group	10	5	4	9	12	6	7

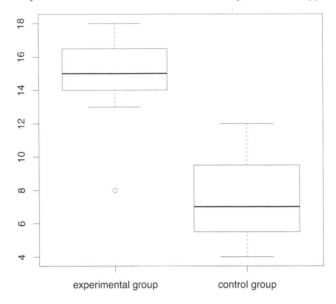

FIGURE 7.1: Box plots for the data of Williams and Carnine (1981) presented in Table 7.1.

significant results at $\alpha = 0.05$. The Lepage test based on mean ranks gives a p-value of 0.0064. When using the modified Lepage test L_M, the smaller p-value 0.0029 results. The Cucconi test is also significant with $C = 3.7928$ and a p-value of 0.0064. However, there is one tied group and, hence, the result of the Cucconi test changes when the groups are interchanged (see Section 3.1). In this example with one tied group only, the difference is small. One would obtain $C = 3.7964$ and a p-value of 0.0061.

When applying a closed testing procedure (see appendix), further tests can follow in a second step because the location-scale test in the first step is significant. Thus, separate tests for differences in location and variability can follow.

For testing differences in location, the Brunner-Munzel test gives $W_{BF} = 6.48$. Based on the exact permutation null distribution, the p-value is 0.0035. Hence, the null hypothesis H_0^{BF}: $p = 1/2$ can be rejected at the significance level $\alpha = 5\%$.

The t test statistic t_{BF} is 4.22; and with 20,000 bootstrap samples, a p-value of 0.0052 was observed. The D.O test proposed by Manly and Francis (2002) also yields a very small p-value: 0.0039. However, the bootstrap validation (with 5,000 bootstrap samples) needed for this test estimated the actual level as 7.2%. This value is significantly larger than 5%. Hence, the D.O test procedure cannot reject the null hypothesis, which illustrates the loss of power of the D.O test caused by the bootstrap validation.

TABLE 7.2: Reaction times (in msec) in two groups of a clinical trial

Placebo	Active Drug
154, 155, 158, 159, 161	171, 172, 178, 179, 184
163, 177, 183, 192, 219	185, 186, 194, 196, 223
Mean: 172.1	Mean: 186.8
Variance: 437.2	Variance: 229.5

Source Sedlmeier and Renkewitz 2008.

In order to test for a difference in variability, the values $\widetilde{Z}_{ij}$ are used. However, there is no evidence for heteroscedasticity. The p-value of the FPP test based on the transformed values is 1.

In summary, there is a difference between the two groups, albeit with regard to location only. Thus, the location-shift model considered in Chapter 2 seems appropriate for this example. The tests discussed in Sections 2.1 through 2.3 give the following results:

FPP test: $P = 24.5$, p = 0.0035,
WMW test: $W = 74$, p = 0.0035,
BWS test: $B = 5.17$, p = 0.0029.

The difference in location can also be observed based on the location-shift model. Consistent with the simulation results (see Section 2.4), the smaller p-value of the BWS test indicates that this test can be advantageous.

7.2 A Clinical Trial

In the clinical trial considered here, twenty patients were randomized to two groups of equal size. In one group an active drug was administered. In the other group the patients received a placebo. The endpoint is the reaction time necessary to react on a visual signal. Raw data are presented in Table 7.2 (according to Sedlmeier and Renkewitz, 2008, p. 583).

As in the previous example we do not have random samples of defined populations. Homogeneous patients were randomized; thus, the randomization model and permutation tests are appropriate again.

Figure 7.2 displays the box plots. One can see differences in both location and variability, consistent with the empirical means and variances given in Table 7.2. Hence, it is no surprise that the location-scale tests are significant at $\alpha = 0.05$. Lepage's test statistic is 6.3455, using the asymptotic null distribution the p-value 0.0419 results. The p-value is 0.0340 when using the permutation

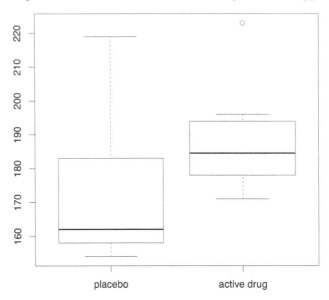

FIGURE 7.2: Box plots for the data of Sedlmeier and Renkewitz (2008) presented in Table 7.2.

null distribution. The modified statistic L_M is 8.3143, with a corresponding (exact) p-value of 0.0329. The Cucconi test is also significant, with $C = 3.0404$ and the p-value 0.0387.

Thus, separate tests for a difference in location and variability, respectively, can follow. The Brunner-Munzel test gives $W_{BF} = 2.45$ and an exact p-value of 0.0345. Thus, the null hypothesis H_0^{BF}: $p = 1/2$ can be rejected at the 5% level. The p-value based on the approximate t distribution is 0.0309. Consistent with this significant result, the estimated relative effect clearly deviates from 0.5: $\hat{p} = 0.78$. The bootstrap test based on t_{BF} and 20,000 bootstrap samples is not significant; the p-value is 0.0959. A p-value of similar size results when applying the Welch t test.

A difference in variability is tested based on the transformed values $\widetilde{Z}_{ij}$. The FPP test applied after this transformation gives a p-value of 0.5135. Although this is far from being significant, the location model seems not to be appropriate for this example because of the heteroscedasticity recognizable in Figure 7.2. However, in order to further illustrate the FPP, WMW, and BWS tests, results are given below. Now, a significance does not give evidence for a pure difference in location, but for a general difference between the two groups.

The result of the WMW test is similar to that of the Brunner-Munzel test: The exact p-value is 0.0355, the asymptotic one is 0.0343. For the BWS test we have $B = 3.0466$ and an exact p-value of 0.0301. Because the sample size is

TABLE 7.3: Soil lead values from two different districts, 20 observations per district

District 1	District 2
16.0, 34.3, 34.6, 57.6, 63.1,	4.7, 10.8, 35.7, 53.1, 75.6,
88.2, 94.2, 111.8, 112.1, 139.0,	105.5, 200.4, 212.8, 212.9, 215.2,
165.6, 176.7, 216.2, 221.1, 276.7,	257.6, 347.4, 461.9, 566.0, 984.0,
362.8, 373.4, 387.1, 442.2, 706.0	1040.0, 1306.0, 1908.0, 3559.0, 21679.0

Source Mielke et al. 1999.

ten per group, and because there are no ties, the asymptotic BWS test may be applied, its p-value is 0.0259. The p-value of the FPP test is distinctly larger: 0.0919—similar to the results obtained with tests based on t_{BF}. The classical t test does not give a significance at $\alpha = 0.05$, either, the p-value being 0.0886.

7.3 Heteroscedasticity

The data set discussed in this section contains soil lead values from two different districts in New Orleans. Berry et al. (2002) analyzed these data; they also presented the raw data given here in Table 7.3. More details on this study can be found in Mielke et al. (1999).

There are some extremely large values in district 2. Moreover, the box plots (Figure 7.3) indicate a difference in variability. The location-scale tests, however, are not significant at $\alpha = 0.05$. Lepage's test statistic is $L = 3.888$; the corresponding asymptotic p-value based on the χ^2 distribution (df = 2) is 0.143. The permutation test gives the p-value 0.148. The results of the modified Lepage test are similar: $L_M = 3.819$, p = 0.114. These two permutation tests are based on 40,000 randomly selected permutations. Due to the nonsignificance of the location-scale test, the closed testing procedure stops. Nevertheless, we present further tests in order to illustrate the methods.

The tests for the nonparametric Behrens-Fisher problem do not give evidence for a difference in location. Brunner and Munzel's test statistic is W_{BF} = 1.43, which gives a p-value of 0.162 based on the t approximation (df = 28.97). The p-value of the corresponding permutation test based on 40,000 permutations is slightly smaller, 0.153, but similar to the p-value of the D.O test, which is 0.14. Because this result is not significant, a bootstrap validation is not required. With the bootstrap test we obtained $t_{BF} = 1.36$ and p = 0.382 based on 20,000 bootstrap samples.

A difference in variability is tested with the FPP test based on the trans-

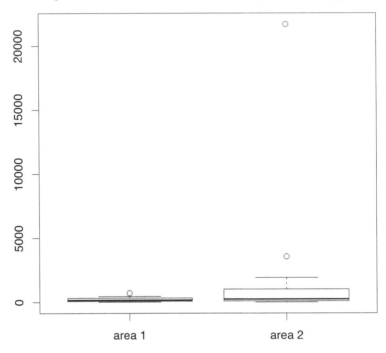

FIGURE 7.3: Box plots for the data from Mielke et al. (1999) presented in Table 7.3.

formed values $\widetilde{Z}_{ij}$; the resulting p-value is 0.0089. However, the t test based on $\widetilde{Z}_{ij}$ is not significant, even when using a significance level of 10% (Neuhäuser, 2007).

In this example we have a significance in the test for a difference in variability although the location-scale tests are not significant. This demonstrates that it is useful to directly test for heteroscedasticity when a potential difference in variability is of primary interest.

Obviously, the location-shift model is not appropriate for this data set with a difference in variability.

7.4   Discrete Numerical Data

Yezerinac et al. (1995) investigated the parentage of nestlings in ninety broods of the yellow warbler (*Dendroica petechia*) using DNA fingerprinting. The yellow warbler is a socially monogamous bird. However, as in many other bird

TABLE 7.4: Observed frequencies of different brood sizes in broods of the yellow warbler

| Broods | Brood Size | | | n_i | Mean | Standard |
	3	4	5			Deviation
Without extra-pair young	10	10	17	37	4.19	0.84
With extra-pair young	4	27	22	53	4.34	0.62

Source Horn 1990.

species, extra-pair fertilizations are common, that is, not all offspring were sired by the territory holder.

Yezerinac et al. (1995) carried out the WMW test in order to compare the brood sizes of nests with and without extra-pair young. This is an extreme example of an application of the WMW test in the presence of ties as the three brood sizes 3, 4, and 5 only occurred. Table 7.4 displays the observed frequencies. The empirical distribution functions are displayed in Figure 7.4. The two functions cross. Hence, the WMW test may have a low power (Horn, 1990). The multitude of ties indicates that one should not apply the asymptotic WMW test.

The two-sample comparison gives the rank sum $W = 2,483$, which is not significant (at $\alpha = 0.05$), either in the asymptotic test (p = 0.52) or in the exact permutation test (p = 0.49). The FPP test is not significant either (p = 0.37). Brunner and Munzel's test statistic for this example is $W_{BF} = 0.6$; the p-value of a permutation test based on 40,000 randomly selected permutations is 0.58. Consistent with the nonsignificant result, the estimate for the relative effect, $\hat{p} = 0.54$, is close to 0.5, the value that holds under the null hypothesis.

Table 7.4 shows that the difference in means is relatively small in this example. Thus, it is no surprise that the three tests mentioned above do not reject their null hypotheses. The BWS test, however, is more sensitive to other differences. For this example we have $B = 16.27$; the corresponding exact p-value is 0.041. Hence, the BWS test shows a significant difference between the broods with and without extra-pair offspring. Relatively small p-values of the Smirnow test (p = 0.066) as well as the χ^2 test (p = 0.014) also indicate that there is some difference between the groups.

7.5 Case-Control Data

As a further example we consider the "German Multicenter Atopy Study" (Liu et al., 2000). This is a case-control study (see Chapter 5) to investigate

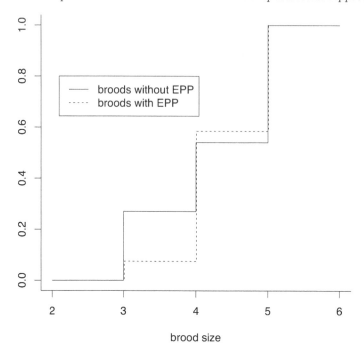

FIGURE 7.4: The empirical distribution functions for the data of Table 7.4 (EPP: extra-pair paternity).

a possible association between atopic dermatitis and a mutation of the *IL13* gene. In this *IL13* gene, a point mutation was detected, the wild-type allele (G) codes for glutamine, the mutant allele (the candidate allele A) for arginine (for further details, see Liu et al., 2000). Table 7.5 shows the genotype distribution of the *IL13* gene for 187 cases with atopic dermatitis and for 98 controls without this disease. In both groups the homozygote wild-type (GG) has the highest frequency.

In the context of such a case-control study for candidate-gene associations, the null hypothesis is the equality of penetrances $f_0 = f_1 = f_2$. A penetrance

TABLE 7.5: Case-control data: The genotype distribution of the *IL13* gene for cases with atopic dermatitis (AD) and controls

	GG	GA	AA
Cases (with AD)	105	72	10
Controls (without AD)	68	24	6

Source Liu et al. 2000.

is the conditional probability of being affected given the respective genotype (see, e.g., Sham, 1998). This null hypothesis is equivalent to the null hypothesis $\pi_1 = \pi_2$ (equality of the multinomial distributions) discussed in Chapter 4 (Freidlin et al., 2002). Here, the one-sided alternative $f_0 \leq f_1 \leq f_2$ with $f_0 < f_2$ is appropriate, because it can be assumed that the homozygote risk is not intermediate between the heterozygote risk and the risk in those without a single copy of the allele A (Sasieni, 1997). Therefore, the modification B^* introduced in Section 2.5 will be applied (Neuhäuser, 2002d).

In this example we have separate random samples of two defined populations: cases and controls. There was no randomization. Hence, the randomization model is not appropriate. Nevertheless, a permutation test might be useful because of the many ties. When applying the Armitage test with equidistant scores (see Chapter 5), the test statistic is $Z = 1.6664$ for this example. Based on the asymptotic standard normal distribution, we get p = 0.0478 (one-sided p-value). The one-sided p-value of the exact permutation test is 0.0581; this is also the p-value of the one-sided FPP test (see Chapter 5).

The Armitage test, also called Cochran-Armitage trend test, can be carried out with the SAS procedure FREQ as follows:

```
DATA ad_study;
   INPUT group result count;
CARDS;
1 1 105
1 2 72
1 3 10
2 1 68
2 2 24
2 3 6
;

PROC FREQ;
   TABLES group*result;
   WEIGHT count;
   EXACT TREND;
RUN;
```

The FPP test can be performed with the SAS procedure NPAR1WAY (see section 2.1). The output of the following program also presents the one-sided p-value 0.0581:

```
PROC NPAR1WAY SCORES=DATA;
   CLASS group;
   VAR result;
```

```
  EXACT;
  FREQ count;
RUN;
```

With the mean ranks $R_1 = \cdots = R_{68} = H_1 = \cdots = H_{105} = 87$, $R_{69} = \cdots = R_{92} = H_{106} = \cdots = H_{177} = 221.5$ and $R_{93} = \cdots = R_{98} = H_{178} = \cdots = H_{187} = 277.5$, we obtain $B^* = 63.52$. The corresponding exact p-value of the one-sided BWS test is 0.0498.

Simulations confirmed that the one-sided exact BWS test can be more powerful than the exact Armitage test for the analysis of case-control data (Neuhäuser, 2002d). Moreover, the BWS test is clearly less conservative. The power comparison between the exact BWS test and the asymptotic Armitage test does not show a clear winner. However, an exact test has the advantage of always guaranteeing the nominal significance level.

The Armitage test is an often-used trend test for binomial data (Corcoran et al., 2000). In that situation the data can also be presented in a 2xk contingency table. However, in that case, k is the number of groups, whereas the rows show the numbers of successes and failures, respectively. In order to apply a trend test (see Section 9.2), the group variable must be ordinal. As an alternative to the Armitage test, a test based on the BWS statistic can also be applied as a trend test for binomial data (Neuhäuser, 2006a). Here, we do not present details because that is not a nonparametric two-sample comparison.

Chapter 8

One-Sample Tests and Tests for Paired Data

Charles Darwin (1876) carried out experiments with corn, *Zea mays*. He used fifteen pairs of seedlings. The seedlings in each pair were of the same age and grew together under identical conditions. One plant of each pair was cross-fertilized, the other plant was self-fertilized. Later, Darwin observed the height of the plants; the raw data are given in Table 8.1 (according to Hand et al., 1994).

The data of this example cannot be permuted in the same way as the data in previous chapters. The observations are paired. The belonging to a pair must be retained; thus, the number of possible permutations is limited. There is only one way to obtain permutations: the values within each pair can be interchanged. If there is no difference in the height between cross- and self-fertilized plants, each of these permutations has the same probability to occur as the actually observed one.

Two permutations are possible per pair; hence, in the example with fifteen pairs there are in total $2^{15} = 32,768$ different permutations. If the data are independent, one could freely permute all thirty observations, which would result in a much higher number of permutations: $\binom{30}{15} \approx 155$ million.

When selecting a test statistic, one should also consider that the observations are paired. The statistics discussed in Chapter 2 do not take into account such a structure within the data and, hence, cannot be suggested. Due to the pairing, a difference between the two values of each pair can be calculated. This difference, which is also given in Table 8.1, can be used to compute a test statistic.

If there is one sample only rather than paired data, the values of the single sample can be analyzed as the differences of paired data. Here, we assume that the differences, or the values of a single sample, respectively, are independent and identically distributed.

TABLE 8.1: Height of corn in an experiment performed by Charles Darwin.

Pair	Cross-fertilized	Self-fertilized	Difference	Rank (with sign)
1	23.5	17.4	6.1	11
2	12.0	20.4	-8.4	-14
3	21.0	20.0	1.0	2
4	22.0	20.0	2.0	4
5	19.1	18.4	0.7	1
6	21.5	18.6	2.9	5
7	22.1	18.6	3.5	7
8	20.4	15.3	5.1	9
9	18.3	16.5	1.8	3
10	21.6	18.0	3.6	8
11	23.3	16.3	7.0	12
12	21.0	18.0	3.0	6
13	22.1	12.8	9.3	15
14	23.0	15.5	7.5	13
15	12.0	18.0	-6.0	-10

Source Hand et al. 1994.

8.1   The Sign Test

If there is no difference between the cross- and self-fertilized offspring, the probability of a positive difference is equal to the probability of a negative difference. This equality of the probabilities is the null hypothesis.

Obviously, one can compare the number of positive and the number of negative differences. Let us first assume that no single difference can take the value 0. In this situation the expected number of positive differences under the null hypothesis is $n/2$, where n is the total number of pairs. The number of positive differences, denoted by N_+, is used as the test statistic. In the example we have $n = 15$ and 13 positive differences.

Now, we list all $2^{15} = 32,768$ possible permutations. They are equiprobable under the null hypothesis. Hence, the permutation null distribution of the test statistic N_+ given in Table 8.2 results. The probabilities displayed in Table 8.2 can also be computed using the binomial distribution with $n = 15$ and a success probability of $p = 0.5$.

The observed value of the test statistic N_+ is $n_+ = 13$. Hence, on the one hand, the probabilities for $N_+ \geq 13$ must be considered for the p-value. On the other hand, the probabilities for $N_+ \leq 2$ must be added in a two-sided test. The expected value of N_+ under the null hypothesis is $n/2$ $(=$

TABLE 8.2: The exact permutation null distribution of the number N_+ of positive differences for the example of Darwin with $n = 15$.

Possible Value of N_+	Probability (= proportion within the 32 768 permutations)
0	0.00003
1	0.00046
2	0.00320
3	0.01389
4	0.04166
5	0.09164
6	0.15274
7	0.19638
8	0.19638
9	0.15274
10	0.09164
11	0.04166
12	0.01389
13	0.00320
14	0.00046
15	0.00003

7.5 in the example). Thus, the values with $N_+ \leq 2$ are as far away from the expected value as the values with $N_+ \geq 13$. Therefore, the two-sided p-value is $2(0.00003 + 0.00046 + 0.00320) = 0.0074$. Of course, this sign test can also be performed one-sided.

Because the probabilities for the different values of N_+ can be computed with the binomial distribution, an exact sign test is possible for relatively large samples. The two-sided p-value can be calculated with the following formula:

$$p - \text{value} = \begin{cases} 2P(N_+ \geq n_+) = 2 \sum_{i=n_+}^{n} \binom{n}{i} 0.5^n & \text{for} \quad n_+ > n/2 \\ 1 & \text{for} \quad n_+ = n/2 \\ 2P(N_+ \leq n_+) = 2 \sum_{i=0}^{n_+} \binom{n}{i} 0.5^n & \text{for} \quad n_+ < n/2, \end{cases}$$

where n_+ denotes the actually observed value of the test statistic.

In the case of large samples, an approximation based on the normal distribution is possible as well. Under the null hypothesis we have $E_0(N_+) = n/2$ and $\text{Var}_0(N_+) = n/4$; thus, $(2N_+ - n)/\sqrt{n}$ is asymptotically standard normal. In the example we have $(2n_+ - n)/\sqrt{n} = 2.84$; the corresponding two-sided asymptotic p-value is 0.0045.

No difference is equal to 0 in the example. When one or more differences are 0, one must consider that not all $n - n_+$ differences are negative. The usually applied approach (see, e.g., Sokal and Rohlf, 1995, p. 444; Hollander and Wolfe, 1999, p. 62) is to ignore all differences that are equal to 0, and to redefine n as the number of differences unequal to 0. This approach is acceptable when the null hypothesis says that the probabilities for a positive and a negative difference are equal, as above.

However, the null hypothesis could also be defined as the statement that the (population) median of the differences is 0. In this case, the zero differences must not be ignored (Larocque and Randles, 2008). If so, one would omit exactly those observations that give evidence for the null hypothesis. If the null hypothesis "median of the difference is 0" is tested against the two-sided alternative "median $\neq 0$," the p-value of a modified sign test according to Fong et al. (2003) can be calculated as follows:

$$P\left(N \geq \max\left(n_-, n_+\right)\right) / P\left(N \geq \left[\frac{n - n_0 + 1}{2}\right]\right).$$

Here, n again denotes the total sample size, N a binomially distributed random variable with the parameters n and $p = 0.5$, n_0 the number of zero differences, and n_- and n_+ the observed numbers of negative and positive differences, respectively. Moreover, let $[x]$ be the floor function, that is, the largest integer that is $\leq x$.

In the example we have $n_0 = 0$ and $\max\left(n_-, n_+\right) = n_+ = 13$. Because $P(N \geq 8) = 0.5$, the exact two-sided p-value is not changed; it is still 0.0074. The usual sign test here is a special case of the modified test proposed by Fong et al. (2003). However, this is only the case when n is odd. If n is even, there is a small difference (Fong et al., 2003).

8.1.1   Implementation in SAS

Because the probabilities given in Table 8.2 can be computed with a binomial distribution, the binomial test can be used to carry out the sign test. The null hypothesis of the binomial test is a statement about the success probability of a binomial distribution, and it is implemented in the SAS procedure FREQ. The following program performs the test for the example with thirteen positive and two negative differences:

```
DATA example8_1;
INPUT diff count;
CARDS;
-1 2
1 13
;
```

```
PROC FREQ;
 TABLES diff;
 WEIGHT count;
 EXACT BINOMIAL;
RUN;
```

In practice, usually a data set with the raw data of Table 8.1 would be used. Then, the numbers of negative and positive differences could be computed using SAS instead of entering the numbers 2 and 13 in the program above. That program produces the following (shortened) output:

```
The FREQ Procedure
```

diff	Frequency	Percent	Cumulative Frequency	Cumulative Percent
-1	2	13.33	2	13.33
1	13	86.67	15	100.00

```
Binomial Proportion for diff = -1
-----------------------------------
Proportion (P)                0.1333

   Test of H0: Proportion = 0.5

Z                            -2.8402
One-sided Pr <  Z             0.0023
Two-sided Pr > |Z|            0.0045

Exact Test
One-sided Pr <=  P            0.0037
Two-sided = 2 * One-sided     0.0074

Sample Size = 15
```

The output contains the estimation for the probability of a negative difference, $2/15 = 0.1333$. By default it is tested in SAS whether or not this probability is equal to 0.5. In the absence of zero differences, this corresponds to the null hypothesis of the equality of the probabilities for positive and negative differences. The exact two-sided p-value is 0.0074, as given above.

Alternatively, the sign test can be carried out using the SAS procedure UNIVARIATE:

```
PROC UNIVARIATE LOCCOUNT;
 VAR difference;
RUN;
```

The variable **difference** must contain the values of the differences. The procedure UNIVARIATE applies not only the sign test, but also the signed rank test (see the following section) and the one-sample t test. The corresponding part of the output is:

```
              Tests for Location: Mu0=0

Test                 -Statistic-     -----p Value------

Student's t     t  2.142152     Pr > |t|     0.0502
Sign            M       5.5      Pr >= |M|    0.0074
Signed Rank     S        36      Pr >= |S|    0.0413


Location Counts: Mu0=0.00

Count                   Value

Num Obs > Mu0            13
Num Obs ^= Mu0          15
Num Obs < Mu0            2
```

The test statistic here is $M = 5.5$, which is defined as $M = 0.5(n_+ - n_-)$. With the option LOCCOUNT, one specifies that the values of n_+, $n-n_0$, and n_- are displayed as "Location Counts." Zero differences are ignored when using this procedure that performs an exact sign test. The p-value is determined based on binomial probabilities.

The aim is not always to distinguish between positive and negative differences. One could define n_+ and n_- as the number of differences that are larger or smaller than, for example, 10. This modified null hypothesis can be tested using the option MU0=10 in the PROC UNIVARIATE statement. By default there is MU0=0.

8.2   The Wilcoxon Signed Rank Test

When applying the sign test we count how many differences are positive. However, the actual values are not incorporated. Now we present a test where the ranks of the differences are entered into the test statistic. At first we again assume that there are no zero differences.

In order to compute the necessary ranks, the signs of the differences are ignored. Thus, the—according to amount—smallest difference gets the rank 1, etc. For the example, these ranks are displayed in Table 8.1. If there are ties, mean ranks can be used. The test statistic is R_+, that is, the sum of the ranks of the positive differences. For the example, the sum R_+ of the positive ranks is 96.

The null hypothesis states that the median of the differences, or the median of the values of one sample, respectively, is 0. For the Wilcoxon signed rank test we must assume that the distribution of the differences is symmetric. Usually, this requirement is fulfilled for differences. Vickers (2005) demonstrated that differences are often symmetric even if computed between skewed variables. Moreover, the difference between two exchangeable variables has a symmetric distribution, as already mentioned in Section 2.4 (Randles and Wolfe, 1979, p. 58).

As in the case of the sign test, all 2^n possible permutations can be considered in order to carry out the Wilcoxon signed rank test in an exact way. In the example with $n = 15$ and 13 positive differences, we have under the null hypothesis $P_0(R_+ \geq 96) = 0.0206$.

When the test is performed two-sided, very small values of R_+ also give evidence against the null hypothesis. The expected value $E_0(R_+)$ is $n(n+1)/4$, that is, 60 in the example. Hence, values of the test statistic that are smaller than or equal to $60 - (96 - 60) = 24$ must be considered for the calculation of the p-value, as all values ≥ 96. Because R_+ has a symmetric distribution (when there are no ties), the two-sided p-value is $P(R_+ \leq 24) + P(R_+ \geq 96) = 2P(R_+ \geq 96) = 0.0413$. If the actually observed value of R_+ is 24, an identical two-sided p-value would result.

The Wilcoxon signed rank test incorporates—in contrast to the sign test—the ranks of the values; hence, it uses more information. Thus, it might astonish that the p-value of the Wilcoxon signed rank test is distinctly larger in the example. The reason is that the two negative differences have a relatively large absolute value, which is ignored by the sign test.

In the case of large samples, an asymptotic p-value can be calculated based on the standard normal distribution. Under the null hypothesis we have $E_0(R_+) = n(n+1)/4$ and $Var_0(R_+) = n(n+1)(2n+1)/24$, or if there are ties,

$$Var_0(R_+) = \frac{1}{24} \cdot \left(n(n+1)(2n+1) - \frac{1}{2} \sum_{j=1}^{g} t_j (t_j - 1)(t_j + 1) \right),$$

where g is the number of tied groups and t_i the number of observations in tied group i. An untied value is considered a "tied group" with size $t_i = 1$ (Hollander and Wolfe, 1999, p. 38). In the example, the standardized test statistic is $(96 - 60)/\sqrt{310} = 2.04$; the resulting two-sided asymptotic p-value is 0.0409.

In the preceding paragraphs, ties relate to the non-zero differences. As mentioned, it was assumed up to now that no difference is 0. In practice zero differences are usually ignored, as in the case of the sign test. An alternative method was introduced by Pratt (1959), who suggested assigning the rank 0 to all zero differences; however, the smallest absolute (i.e., according to amount) non-zero difference gets the rank $n_0 + 1$ rather than 1, etc. The largest absolute value eventually gets the rank n. Here, the number of zero differences is denoted by n_0, as above. In the example data set of Charles Darwin, there are no zero differences; thus the p-value does not change due to Pratt's modification.

Pratt's modification (Pratt, 1959) changes the expected value of the test statistic: instead of $n(n + 1)/4$, we have (Buck, 1979)

$$\frac{n(n + 1) - n_0(n_0 + 1)}{4} \ .$$

The standardized test statistic is still asymptotically standard normal under the null hypothesis (Buck, 1979).

As already mentioned in Section 8.1, the Wilcoxon signed rank test can be performed with the SAS procedure UNIVARIATE. This procedure computes an exact p-value for sample sizes of $n \leq 20$; for larger samples, an approximation based on a t distribution is used. The test statistic S given in the SAS output is defined by $S = R_+ - (n - n_0)(n - n_0 + 1)/4$.

With the SAS procedure UNIVARIATE and the option MU0=, the null hypothesis that the median of the differences is μ_0, is tested. Alternatively, one can perform the test as described without the option MU0, when the value μ_0 is subtracted from each of the differences before the test is carried out (Hollander and Wolfe, 1999, p. 42).

An SAS program that performs the exact Wilcoxon signed rank test also for $n > 20$ can be found at www.egms.de/static/en/journals/mibe/2010-6/mibe000104.shtml (Leuchs and Neuhäuser, 2010). This SAS program (see also Section 8.3) can carry out Pratt's modification too. Obviously, for very large samples, an approximate permutation test can also be applied in the one-sample case.

8.2.1   Comparison between the Sign Test and the Wilcoxon Signed Rank Test

The Wilcoxon signed rank test needs an additional assumption, the symmetry of the distribution. On the other hand, this test utilizes the ranks of the differences. Nevertheless, the signed rank test is not always more powerful than the sign test when the symmetry assumption is justified. The asymptotic relative efficiency shows that there are favorable situations for both tests. For example, the sign test is less efficient in the case of a normal distribution, with an asymptotic relative efficiency of the sign test to the Wilcoxon signed rank test of 2/3. For some other distributions, this asymptotic relative efficiency is larger than 1, which means that the sign test is advantageous. An example is the Cauchy distribution with an asymptotic relative efficiency of 1.3 (Higgins, 2004, p. 127).

8.3   A Permutation Test with Original Observations

Comparable to the FPP test in the two-sample situation, a permutation test can be performed with the original observations of one sample, or with the actual values of the differences, respectively. Instead of using the rank sum of the positive differences, the sum of the positive differences is used as the test statistic. In the example (see Table 8.1), the observed value of this test statistic is $6.1 + 1.0 + \cdots + 7.5 = 53.5$.

Again, one must consider all $2^{15} = 32,768$ permutations. The two-sided exact p-value is 0.0529. In contrast to both the sign test and the signed rank test, the permutation test with original observations is not significant at $\alpha = 0.05$. The reason is that the two negative differences have relatively large absolute values.

As in the Wilcoxon signed rank test, the permutation test with original observations tests the null hypothesis that the median of the differences is 0. Again, it must be assumed that the distribution of the differences is symmetric. The test is implemented in StatXact.

8.3.1   Implementation in SAS

The permutation test with original observations as well as the Wilcoxon signed rank test and its modification according to Pratt (1959) can be carried out with the SAS program freely available at www.egms.de/static/en/journals/ mibe/2010-6/mibe000104.shtml, which is described in detail by Leuchs and Neuhäuser (2010). This SAS macro uses an algorithm proposed by Munzel and Brunner (2002). Here, we present two examples of how the SAS macro can be used to analyze Charles Darwin's data. First, the data must be entered into a SAS data set:

```
DATA darwin;
```

```
INPUT difference @@;
CARDS;
6.1 -8.4 1.0 2.0 0.7 2.9 3.5 5.1 1.8 3.6 7.0 3.0 9.3 7.5 -6.0
;
RUN;
```

In the first example, the permutation test with original observations should be carried out, and all p-values, that is, the two-sided and both one-sided p-values, should be computed. Then, the macro is invoked as follows:

```
%signedrank(darwin, difference, 'original', 'all');
```

The output is:

test	n	n (notnull)	statistic	p-value (less)	p-value (greater)	p-value (two-sided)
original	15	15	53.5	0.9742	0.0264	0.0529

In the second example, the modification of Pratt (1959) should be used, and the two-sided p-value should be displayed:

```
%signedrank(darwin, difference, 'pratt', 'two');
```

In this case, the output is:

test	n	n (notnull)	statistic	p-value (two-sided)
pratt	15	15	96	0.0412598

8.4   One-Sample Bootstrap Tests

The bootstrap can also be used in the one-sample situation. Let us assume that we have one sample $x_1, \ldots, x_n$ of size n, the underlying distribution of which is unknown. Of course, this sample could contain differences. The null hypothesis is that the expected value μ of the differences, or the sample values, respectively, is μ_0. In the case of a symmetric distribution, the median is also μ_0 under this null hypothesis.

A possible test statistic is the one-sample t statistic, $t = \sqrt{n}(\bar{x} - \mu_0)/S$, where $\bar{x}$ denotes the sample mean and $S^2 = \sum_{i=1}^{n} (x_i - \bar{x})^2/(n-1)$ is the sample variance. Because the distribution of t under the null hypothesis $\mu = \mu_0$ is required, the data are transformed in order to get a sample mean of μ_0. To be precise, the following transformation is applied: $\widetilde{x}_i = x_i - \bar{x} + \mu_0$. Then B independent bootstrap samples can be drawn with replacement from the transformed values $\widetilde{x}_1, \ldots, \widetilde{x}_n$. The test statistic must be calculated for every bootstrap sample in order to estimate the null distribution.

With the data of Charles Darwin (see Table 8.1) we get $t = 2.14$ for $\mu_0 = 0$, and with 20,000 bootstrap samples the two-sided p-value is 0.0744. This test can be carried out by means of the SAS procedure MULTTEST, similar to the use of this procedure presented in Section 3.3:

```
DATA darwin;
 INPUT difference dummy1 @@;
 dummy2=1;
CARDS;
6.1 1 -8.4 1 1.0 1 2.0 1 0.7 1 2.9 1 3.5 1 5.1 2 1.8 2 3.6 2
7.0 2 3.0 2 9.3 2 7.5 2 -6.0 2
;
RUN;

PROC MEANS MEAN NOPRINT;
 VAR difference;
 OUTPUT OUT=m MEAN=meanva;
RUN;

DATA _null_;
 SET m;
 CALL SYMPUT ('meanva',meanva);
RUN;

DATA all1;
 SET darwin;
 trans_x=difference-&meanva;
RUN;

PROC MULTTEST DATA=all1 BOOTSTRAP NOCENTER N=20000
 OUTSAMP=all2 NOPRINT;
 CLASS dummy1;
 TEST MEAN(trans_x);
 CONTRAST "1 vs. 2" -1 1;
RUN;

*Computation of the one-sample t test statistic for the
 bootstrap samples;
```

```
ODS OUTPUT TTESTS=s3;
ODS LISTING CLOSE;
PROC TTEST DATA=all2;
 BY _sample_;
 VAR trans_x;
RUN;
ODS LISTING;

*One-sample t test for the original data;
ODS OUTPUT TTESTS=origva;
PROC TTEST DATA=darwin;
 VAR difference;
RUN;

DATA _null_;
 SET origva;
 CALL SYMPUT ('t_orig',tvalue);
RUN;

DATA all4;
 SET s3;
 t_orig=&t_orig;
RUN;

DATA all4;
 SET all4;
 p_value=(ABS(tvalue)>ABS(t_orig))/20000;
RUN;

PROC MEANS SUM DATA=all4;
 VAR p_value;
RUN;
```

Of course, test statistics other than the t statistic could be used in a bootstrap test. Moreover, other null hypotheses can be tested. Examples include a test for the variance (Good, 2001, p. 77) and a test for multimodality (Efron and Tibshirani, 1993, pp. 227–232). In the latter test, we test whether a distribution has only one mode, that is, one maximum, or more than one.

TABLE 8.3: The smoking habits of 300 people before and after a health program

		Smoker Before		Total
		No	Yes	
Smoker	No	132	49	181
After	Yes	21	98	119
	Total	153	147	300

Source Duller 2008.

8.5 The McNemar Test

When there are paired dichotomous data, positive and negative differences cannot differ in their absolute value. The sign test can be applied in this case. Let us consider an example: 300 people took part at a health program, and the aim is to investigate whether or not the program influenced smoking habits. Table 8.3 shows how many people were smokers before and after the program (Duller, 2008, p. 206).

In total, there are 70 people with a change during the health program. These 70 people are unevenly distributed: there are 49 people in the category "yes-no," but only 21 persons in the category "no-yes" (however, these 21 people who are smokers after, but not before, the health program might be too much to recommend this program). When the 230 people without a change (zero differences) are ignored, the two-sided exact p-value of the sign test is 0.0011.

An alternative way to analyze this data example is a χ^2 test. When computing the expected frequencies, one must incorporate the fact that these are paired data. As a consequence, the expected frequencies for the categories "yes-yes" and "no-no" are exactly equal to their observed frequencies. Hence, the test statistic is reduced to the following sum of only two ratios (using the example data):

$$\frac{(49 - 70/2)^2}{70/2} + \frac{(21 - 70/2)^2}{70/2} = \frac{(49 - 21)^2}{49 + 21} = 11.2.$$

This test statistic is known as McNemar test. In general, McNemar's test statistic is $X^2 = (b - c)^2/(b + c)$, where b and c are defined by the following 2x2 table:

	$X = 0$	$X = 1$
$Y = 0$	a	b
$Y = 1$	c	d

Lu (2010) introduced the following modification of McNemar's statistic that also incorporates a and d:

$$\frac{(b-c)^2}{(b+c)\left(1+\frac{a+d}{n}\right)},$$

where n is the sample size. Lu (2010) recommended using the new statistic when $b+c \geq 10$.

8.5.1   Implementation in SAS

The usual exact McNemar test can be performed using the SAS procedure FREQ as follows:

```
DATA smoking;
 INPUT before $ after $ count;
 CARDS;
 no no 132
 no yes 21
 yes no 49
 yes yes 98
 ;
RUN;

PROC FREQ;
 TABLES before*after;
 WEIGHT count;
 EXACT MCNEM;
RUN;
```

The output presents the test statistic with the value 11.2, the asymptotic p-value 0.0008, and the exact p-value 0.0011. Because this test corresponds to the sign test, it can also be carried out using the SAS procedure UNIVARIATE. In order to apply that procedure, one needs numerical codes for the variables "before" and "after". The reason is that a difference such as "before – after" must be calculated. Moreover, the PROC UNIVARIATE output presents another test statistic: $M = 0.5(n_+ - n_-) = 0.5(49 - 21) = 14$ (see Section 8.1) instead of $X^2 = 11.2$.

Chapter 9

Tests for More than Two Groups

9.1 The Kruskal–Wallis Test and the Permutation F Test

The principles of permutation and bootstrap tests can also be applied in the case of more than two groups. In the simplest case of a completely randomized design with k independent groups, the permutations are the possible allocations of the observations to the k groups. Possible test statistics include the rank-based Kruskal-Wallis statistic and the F statistic of the one-way analysis of variance (ANOVA). By analogy with the two-sample Fisher-Pitman permutation test (section 2.1) the F statistic can be computed for every permutation, and the resulting permutation null distribution can be used for inference instead of the F distribution. For this permutation test there are other equivalent test statistics. For example, the mean square between the groups, that is, the numerator of the F statistic, can be used as the test statistic (Manly, 2007, p. 136).

When k, the number of groups, increases, the number of possible permutations increases even faster. In this case a permutation test can be performed approximately based on a random sample of permutations. Moreover, there are algorithms such as the network algorithm of Mehta and Patel (1983), which clearly need less computing time than any "naive" counting of permutations.

Here, we consider the following model: The sample size in group i is denoted by n_i, $i = 1, \ldots, k$. The total sample size of all k groups is N. Then, the number of possible permutations is $N! / \prod_{i=1}^{k} n_i!$. The observations X_{ij} within the k groups are independent, and distributed according to distribution function F_i, that is:

$$X_{ij} \sim F_i, \quad i = 1, 2, \ldots, k, \ j = 1, 2, \ldots, n_i.$$

Often, as in Chapter 2 for the two-sample case, it is assumed within a location-shift model that the distributions of the different groups differ, if at all, by a shift in location only. Thus, the functions $F_1, \ldots, F_k$ have the same shape, but can differ with respect to a location parameter ϑ_i:

$$F_i(t) = F(t - \vartheta_i),$$

where F is a distribution function. The null hypothesis is the equality of the distributions of all k groups; hence, there is no difference in location:

$$H_0 : \vartheta_1 = \vartheta_2 = \cdots = \vartheta_k .$$

Under the alternative there is at least one difference between two, or more, groups: Formally, the alternative is "there is a pair i, j with $\vartheta_i \neq \vartheta_j$." In this situation the Kruskal-Wallis test can be applied; its test statistic H is defined as follows (see, e.g., Duller, 2008, p. 215):

$$
\begin{aligned}
H &= \frac{12}{N(N+1)} \sum_{i=1}^{k} \frac{1}{n_i} \left(R_i - \frac{n_i(N+1)}{2} \right)^2 \\
&= \left(\frac{12}{N(N+1)} \sum_{i=1}^{k} \frac{R_i^2}{n_i} \right) - 3(N+1),
\end{aligned}
$$

where R_i is the rank sum in group i, $i = 1, \ldots, k$. All N observations are pooled for ranking.

Under the null hypothesis, the rank sum R_i has the expected value $n_i(N+1)/2$. Thus, the Kruskal-Wallis test statistic cumulates the differences between the rank sums and their expected values. Because these differences are squared, the Kruskal-Wallis test is two-sided.

We do not assume continuous distributions; hence, ties are possible. The statistic H is modified as follows in the case of ties (see, e.g., Duller, 2008, p. 215):

$$H^* = \frac{H}{C} \text{ with } C = 1 - \frac{\sum_{i=1}^{g} (t_i^3 - t_i)}{N^3 - N} .$$

As in Chapter 2, g denotes the number of tied groups and t_i the number of observations in tied group i. An untied observation is considered a tied group of size $t_i = 1$.

With the test statistic H^* an exact permutation test, an approximate permutation test, or an asymptotic test can be performed. The latter is possible because H^* is, for large sample sizes n_i, approximatively χ^2 distributed with $k - 1$ degrees of freedom. Because C is constant for all permutations, a permutation test is possible with H even in the case of ties.

Let us consider an example presented by Zöfel (1992): Persons in four groups were differently motivated, for example, with a compliment or cash. The level of concentration was measured. In the control group there was no special motivation. Table 9.1 shows the obtained points.

TABLE 9.1: Points obtained in a test of level of concentration in groups with different motivations

Group	Points						
A (Control group)	9	11	7	8	9	9	10
B	11	12	8	9	10	11	10
C	12	13	11	9	10	12	12
D	15	17	15	10	16	14	12

Source Zöfel 1992.

First, we need the ranks of all $N = 28$ observations. For ties, mean ranks are used. Hence, the ranks in group A are 6, 15.5, 1, 2.5, 6, 6, and 11; the rank sum is 48. The other three rank sums are 81.5, 115.5, and 161. Therefore, the test statistic has the value $H = 14.78$.

In this example, there are many ties. Only five out of twenty-eight obervations are untied. Nevertheless, the statistic C does not differ much from 1, and we have $C = 0.98$. Thus, $H^* = 14.78/0.98 = 15.07$. Based on the χ^2 distribution with $k - 1 = 3$ degrees of freedom, 0.0018 is the asymptotic p-value. However, a permutation test seems preferable because of the small samples and the multitude of ties. In total there are $N!/ \prod_{i=1}^{k} n_i! = 28!/7!^4 = 472 \cdot 10^{12}$ permutations. Despite this huge number of permutations, an exact permutation test is possible with the help of efficient algorithms, the resulting p-value is 0.0003. Of course, an approximate permutation test would be an alternative in this case.

9.1.1 Implementation in SAS

The Kruskal-Wallis test can be applied as the Wilcoxon-Mann-Whitney (WMW) test (see Section 2.2) with the SAS procedure NPAR1WAY and the option WILCOXON:

```
PROC NPAR1WAY WILCOXON;
 CLASS group;
 VAR point;
 EXACT;
RUN;
```

Now, the CLASS variable has more than two categories. The statement EXACT invokes the Kruskal-Wallis test as an exact permutation test. The output displays the Wilcoxon scores (= rank sums) of the different groups as well as the result of the Kruskal-Wallis test, both asymptotically and exact:

The NPAR1WAY Procedure

Wilcoxon Scores (Rank Sums) for Variable point
          Classified by Variable group

| | | Sum of | Expected | Std Dev | Mean |
group	N	Scores	Under H0	Under H0	Score
1	7	48.00	101.50	18.661458	6.857143
2	7	81.50	101.50	18.661458	11.642857
3	7	115.50	101.50	18.661458	16.500000
4	7	161.00	101.50	18.661458	23.000000

Average scores were used for ties.

Kruskal-Wallis Test

Chi-Square		15.0721
DF		3
Asymptotic Pr > Chi-Square		0.0018
Exact Pr >= Chi-Square		2.807E-04

When the option WILCOXON is replaced by SCORES=DATA, a test based on the actual values is performed rather than a rank test. In SAS the F statistic mentioned above is not used. Instead, the following statistic T is applied:

$$T = \frac{1}{S^2} \sum_{i=1}^{k} n_i (\bar{X}_i - \bar{X})^2 \, ,$$

where $\bar{X}_i$ denotes the sample mean of group i and $\bar{X}$ the mean of the pooled sample. Moreover,

$$S^2 = \frac{1}{N-1} \sum_{i=1}^{k} \sum_{j=1}^{n_i} (X_{ij} - \bar{X})^2 \, .$$

The statistic T is asymptotically χ^2 distributed with $k-1$ degrees of freedom, under the null hypothesis that there is no difference. Thus, the asymptotic test differs from the F test. However, the permutation test with T is equivalent to the permutation test with the F statistic (Cytel, 2007, p. 358). This test is called the Fisher-Pitman permutation test also in case of more than two groups (Neuhäuser and Manly, 2004).

For the example of Table 9.1 we have $T = 16.30$, the asymptotic p-value based on the χ^2 distribution with df = 3 is 0.0010, the exact permutation test gives the p-value 0.0001 (rounded to four decimal places), as shown by the following SAS output:

```
    Data Scores One-Way Analysis

Chi-Square                       16.3002
DF                                     3
Asymptotic Pr >  Chi-Square       0.0010
Exact      Pr >= Chi-Square     8.306E-05
```

If the null hypothesis of equality of all k groups can be rejected, often the question arises as to which groups differ. Then, the two-sample tests presented in the previous chapters can be used for the necessary comparisons of pairs of groups.

There are many further tests for the comparison of k independent samples. Obviously, bootstrap tests are also possible in this situation (see, e.g., Wilcox, 2003, pp. 309–314). Moreover, one can relax the assumptions of the location-shift model. For instance, a location-scale test, described in Section 3.1 for the two-sample case, is possible. Lepage's statistic can be generalized for more than two groups, for example, by combining the Kruskal-Wallis statistic with a generalized Ansari-Bradley test; see Murakami (2008) for details. Furthermore, there is an alternative to the Kruskal-Wallis test that tests for a difference in location in case of potentially unequal variances, as the Brunner-Munzel test (see Section 3.2) does in the two-sample situation. The test for more than two groups is based on a method introduced by Brunner et al. (1997), and it is described in detail by Wilcox (2003, pp. 568–571).

9.2 Trend Tests

The tests discussed above in Section 9.1 can detect arbitrary location differences between the groups. When, however, the grouping variable has an (at least) ordinal scale, it is often appropriate to specify the alternative more precisely. This is the case, for example, when the groups differ in the dose of a treatment and when it can be assumed that the effect of the treatment does not decline with increasing dose.

Let us consider an example. The Ames assay is a test for mutagenicity applied in order to assess the mutagenic potential of chemical compounds (Göggelmann, 1993). This assay can recognize chemically induced mutagenic—and often accompanied carcinogenic—effects. Specific strains of bacteria are used that, due to a gene mutation, require histidine for growth. These bacteria can mutate back to the wild-type, especially when exposed

TABLE 9.2: Number of observed revertant colonies in an Ames assay

Dose (in μg)	Count		
0	101	117	111
10	91	90	107
30	103	133	121
100	136	140	144
300	190	161	201
1000	146	120	116

Source From Hothern 1990.

to a mutagenic compound. The revertants are able to grow on a histidine-free medium. The variable of interest is the number of revertant colonies per plate. The distribution of this counting variable is not clear; nonparametric tests are appropriate. The example data displayed in Table 9.2 were presented by Hothorn (1990, p. 201); three plates were investigated per dose.

The dose is a quantitative variable. Hence, one could use the values of this variable within, for example, a regression analysis. However, the dose is often treated as an ordered categorical variable. This "ANOVA approach uses dose levels as discrete variables and is generally more suitable when few doses are studied" (Ruberg, 1995, p. 16). Note that this approach does not require the selection of a specific model.

As mentioned in Section 9.1, the largest possible alternative "there is a pair i, j with $\vartheta_i \neq \vartheta_j$" states that there is any difference in location. This alternative is not appropriate when the question is whether or not a trend exists. When applying a trend test, the alternative hypothesis is constrained to be one-sided and ordered:

$$\mathrm{H}_1^T : \vartheta_1 \leq \vartheta_2 \leq \cdots \leq \vartheta_k \quad \text{with} \quad \vartheta_1 < \vartheta_k .$$

Such an ordered alternative is appropriate when different doses of the same compound were investigated—as in the example presented in Table 9.2. Of course, the alternative hypothesis can also be formulated for a decreasing trend: $\vartheta_1 \geq \cdots \geq \vartheta_k$ with $\vartheta_1 > \vartheta_k$. We do not consider this case; it is completely analogous to the increasing trend. One can analyze $-X_{ij}$ rather than X_{ij}, due to the difference in the sign of the direction of the trend changes. One can also formulate a two-sided trend alternative; however, it is usually of no relevance for statistical practice.

The standard nonparametric method for an ordered alternative is the trend test proposed by Jonckheere (1954) and Terpstra (1952). The test statistic is a sum of Mann-Whitney scores. The Mann-Whitney score with regard to the groups i and j is defined as

$$U_{ij} = \sum_{t=1}^{n_i} \sum_{s=1}^{n_j} \phi(Y_{it}, Y_{js}) \quad \text{with} \quad \phi(a,b) = \begin{cases} 1 & \text{if} \quad a < b \\ 0.5 & \text{if} \quad a = b \\ 0 & \text{if} \quad a > b. \end{cases}$$

For ties, ϕ is set to $1/2$, as suggested by Hollander and Wolfe (1999, p. 203). The test statistic of the Jonckheere-Terpstra test is

$$T_{JT} = \sum_{i=1}^{k-1} \sum_{j=i+1}^{k} U_{ij} .$$

In the case of large values of the test statistic, this test rejects the null hypothesis in favor of H_1^T. Based on T_{JT}, an exact as well as an approximate permutation test can be carried out. In the case of large sample sizes, one can also use the asymptotic normality of the test statistic. To be precise, with $N = \sum_{i=1}^{k} n_i$, the standardized statistic

$$\frac{\left(T_{JT} - \dfrac{N^2 - \sum_{i=1}^{k} n_i^2}{4} \right)}{\sqrt{\mathrm{Var}_0(T_{JT})}}$$

$$\text{with} \quad \mathrm{Var}_0(T_{JT}) = \frac{N^2(2N+3) - \sum_{i=1}^{k} n_i^2(2n_i+3)}{72}$$

is, under the null hypothesis, asymptotically (i.e., for $\min\{n_1, n_2, \ldots, n_k\} \to \infty$) standard normal (Hollander and Wolfe, 1999, p. 203). Thus, the null hypothesis is rejected in an asymptotic test at level α if the standardized statistic is larger than the $(1 - \alpha)$ quantile of the standard normal distribution.

The formula given above for the standardization of T_{JT} only holds when there are no ties. In the presence of ties, the variance is reduced to

$$\mathrm{Var}_0(T_{JT}) = \frac{1}{72}[N(N-1)(2N+5)$$

$$- \sum_{i=1}^{k} n_i(n_i - 1)(2n_i + 5) - \sum_{j=1}^{g} t_j(t_j - 1)(2t_j + 5)]$$

$$+ \frac{1}{36N(N-1)(N-2)} \left(\sum_{i=1}^{k} n_i(n_i - 1)(n_i - 2) \right) \left(\sum_{j=1}^{g} t_j(t_j - 1)(t_j - 2) \right)$$

$$+ \frac{1}{8N(N-1)} \left(\sum_{i=1}^{k} n_i(n_i - 1) \right) \left(\sum_{j=1}^{g} t_j(t_j - 1) \right) .$$

Again, g denotes the number of tied groups and t_i the number of observations in the tied group i. If there are no ties, we have $t_i = 1$ for all i, and both mentioned formulas to calculate the variance give an identical result (Hollander and Wolfe, 1999, p. 212).

For the example data of Table 9.2 we get $T_{JT} = 104$, and the standardized test statistic is $(104 - 67.5)/12.99 = 2.81$. The resulting asymptotic one-sided p-value is 0.0025, however, the sample sizes $n_i = 3$ are too small to apply an asymptotic test. The exact one-sided p-value of the permutation test is $P_0(T_{JT} \geq 104) = 0.0022$.

After a significant trend test, further tests are possible. However, because of the order restriction it does not make sense to perform all possible pair comparisons. Instead, the following procedure of Lüdin (1985) can be used in order to estimate the minimum effective dose: First, the Jonckheere-Terpstra test is performed with all groups. If this test is significant, the highest dose is proved to be effective, and the test is repeated without this highest dose. If this second trend test is significant as well, the second highest dose is effective, too. Then, this dose is also omitted for the third test, etc. In case of a non-significant test, the procedure stops. The highest dose in the last significant trend test is the estimated minimum effective dose. Since this method is a closed testing procedure (see appendix) every single test can be performed with the full level α. If the test has to be carried out with only two doses, the one-sided WMW test is applied as a special case of the Jonckheere-Terpstra test.

In the example the following three exact Jonckheere-Terpstra tests are significant at the 5% level: the test with all six groups of Table 9.2 (one-sided p-value: 0.0022), the test with the doses up to 300 μg (p-value: 0.0001), and the test with the doses up to 100 μg (p-value: 0.0123). The test with the doses 0, 10, and 30 μg is not significant (p-value: 0.3339), hence, the procedure stops. The efficacy (mutagenicity in this example) can be shown up to the dose 100 μg. However, one should consider that the sample sizes and therefore the power of the tests are low. With larger samples the test with the three groups up to 30 μg might be significant.

9.2.1 Implementation in SAS

The Jonckheere-Terpstra test can be carried out with the SAS procedure FREQ:

```
DATA ames;
 INPUT dose count @@;
 CARDS;
0 101 0 117 0 111
10 91 10 90 10 107
```

```
30 103 30 133 30 121
100 136 100 140 100 144
300 190 300 161 300 201
1000 146 1000 120 1000 116
;
RUN;

PROC FREQ;
 TABLES dose*count;
 EXACT JT;
RUN;
```

The statement EXACT with the option JT invokes the permutation test with the test statistic T_{JT}. The asymptotic test is applied in addition, as the following output shows:

```
Statistics for Table of dose by count


Jonckheere-Terpstra Test

Statistic (JT)              104.0000
Z                             2.8098


Asymptotic Test
One-sided Pr >  Z             0.0025
Two-sided Pr > |Z|            0.0050


Exact Test
One-sided Pr >=  JT           0.0022
Two-sided Pr >= |JT - Mean|   0.0043


Sample Size = 18
```

9.2.2   Comparison of the Jonckheere–Terpstra Test with Other Tests

As expected, the Jonckheere-Terpstra test is more powerful than the Kruskal-Wallis test for monotonically increasing trends $\vartheta_1 \leq \vartheta_2 \leq \cdots \leq \vartheta_k$ (Magel, 1986). However, there are a lot of further nonparametric trend tests. Mahrer and Magel (1995) compared the Jonckheere-Terpstra test with the tests of Cuzick (1985) and Le (1988) and did not find any major differences. Mahrer and Magel (1995, p. 870) wrote: "the tests yielded similar powers for the detection of the trends under the alternative hypothesis. Since the powers are so close among the three tests, it seems reasonable to recommend that researchers employ the test they find easiest to use." Therefore, we focus on the Jonckheere-Terpstra test, which is the most often applied nonparametric trend test (Budde and Bauer, 1989). Furthermore, Weller and Ryan (1998) recommended this test for the analysis of count data.

In the case of small samples, it can be useful to apply the modified Jonckheere-Terpstra statistic

$$T_{MJT} = \sum_{i=1}^{k-1} \sum_{j=i+1}^{k} (j-i)\, U_{ij} \, ,$$

because it can be less discrete (Neuhäuser et al., 1998).

Another rank-based trend test statistic was introduced by Hettmansperger and Norton (1987). That test is described by Brunner and Munzel (2002, pp. 118–119) in addition to the Jonckheere-Terpstra statistic. Alternatively, one can apply a maximum test when no single test statistic can be selected; see Neuhäuser et al. (2000) for details.

For binary data, the Armitage test described in Chapter 5 is usually the applied trend test (Corcoran et al., 2000). However, is is also possible to apply the principle of the Jonckheere-Terpstra test for dichotomous data (Neuhäuser and Hothorn, 1998). Moreover, the modified BWS statistic B^* (see Section 2.5) can also be used in order to construct a trend test for binomially distributed data (Neuhäuser, 2006a).

9.2.3   Tests for Umbrella Alternatives

Hothorn (1990, p. 201) also presented a further dose not included in Table 9.2. It is the dose 3,000 μg with the observed counts 92, 102, and 104. The number of revertants seems to decline at very high doses. This number first increases and then decreases:

$$\vartheta_1 \leq \vartheta_2 \leq \cdots \leq \vartheta_u \geq \vartheta_{u+1} \geq \cdots \geq \vartheta_k$$

$$\text{with} \quad \vartheta_1 < \vartheta_u > \vartheta_k \, .$$

Such a pattern is called an *umbrella alternative*. It can occur, for example in an Ames assay, when the investigated compound is not only mutagenic, but also toxic at high doses. Due to the toxicity, even revertants cannot form colonies.

When the values of the high dose 3,000 μg are included, the Jonckheere-Terpstra test is no longer significant, with an exact one-sided p-value of 0.1376. The umbrella pattern seems to start near the dose 1,000 μg; the mean count is 127.3 for that dose compared to 184.0 for the dose 300 μg. This decrease is relatively low, so that the significance in the Jonckheere-Terpstra test is not lost (see above). This significance, however, is stronger with an exact one-sided p-value of 0.0001 when only the doses 0 to 300 μg are included.

There are specific tests for umbrella alternatives (Chen, 1991). If the peak u is known, the following test statistic might be appropriate:

$$T_u = \sum_{i=1}^{u-1} \sum_{j=i+1}^{u} U_{ij} + \sum_{i=u}^{k-1} \sum_{j=i+1}^{k} U_{ji} \, .$$

For the doses up to the peak, the usual Jonckheere-Terpstra statistic is used. After the peak, the statistic U_{ji} instead of U_{ij} is applied. The peak, however, is hardly ever known in practice. For an unknown peak, the statistic T_u can be computed for all possible values of u, and the maximum is used as the test statistic:

$$T_{CW} = \max(T_1, \ldots, T_k).$$

A permutation test can be performed with this statistic T_{CW} proposed by Chen and Wolfe (1990). An asymptotic test is also possible as T_{CW} is asymptotically normal (Chen, 1991). Formulas for the expected value and the variance needed for the standardization of T_{CW} are given by Chen (1991).

The tests based on the statistics T_u or T_{CW} have one serious disadvantage: They can give a significance when there is no increase up to dose u. The decrease after u can be sufficient to reject the null hypothesis. In an extreme case, $u = 1$ is possible so that there is a decrease without any previous increase. This case $\vartheta_1 \geq \cdots \geq \vartheta_k$ with $\vartheta_1 > \vartheta_k$ is part of the usual umbrella alternative that is often not appropriate, for instance, when the decrease in an Ames assay is caused by the toxicity of high doses. Another example is an agricultural field experiment when the yield decreases due to overfertilization at high doses of fertilizer.

A modified umbrella alternative can be defined for $u \leq k$ as follows:

$$\vartheta_1 \leq \vartheta_2 \leq \cdots \leq \vartheta_u \quad \text{with} \quad \vartheta_1 < \vartheta_u \, , \quad u \geq 2.$$

Here, the pattern of the groups $u + 1$ to k is of no relevance. It is crucial whether or not there is an initial trend up to the group u. After that group, there is no restriction. This test problem is called "protected trend alternative" (Bretz and Hothorn, 2001). The test statistic can be defined as follows:

$$T_{PT} = \max(\widetilde{T}_2, \ldots, \widetilde{T}_k).$$

$$\text{with} \quad \widetilde{T}_u = \sum_{i=1}^{u-1} \sum_{j=i+1}^{u} U_{ij} \, .$$

In T_{PT} there is no statistic $\widetilde{T}_1$ because at least an increase from dose 1 to dose 2 is required. Further restrictions are also possible; the maximum in T_{PT} can be over the domain from f to g, instead 2 to k; in that case, $2 \leq f \leq u \leq g \leq k$ is required.

Obviously, a permutation test can be carried out with T_{PT}. For an asymptotic test, the expected value and the variance of T_{PT} are needed. Formulas can be deduced from the results given by Tryon and Hettmansperger (1973) and Chen (1991).

9.3   Tests for Multiple Dependent Groups

Now we assume that there are k dependent groups. For example, k different treatments are given to every patient. In that case, every patient would be a block. A block could also consist of a homogeneous group of persons, or experimental units. Ideally, the order of the different treatments is randomized within each block. Moreover, when there is one observation of each of the k treatments in each of the n blocks, the design is called a randomized complete block design. As in the case of two paired groups (see Chapter 8), not all permutations are possible. Separately for each block, the k values can be permuted.

For a rank test, the ranks are computed separately for each block, and the rank sum R_i can be calculated for each treatment group. The total sum of ranks for all n blocks and all k groups is $nk(k+1)/2$. Under the null hypothesis, there is no difference between the groups, and the expected value of the rank sum is $n(k + 1)/2$ for every group. For each group, the difference between the rank sum and its expected value can be computed, and squared. The test statistic of the Friedman test is, sum of these squares (see, e.g., Hollander and Wolfe, 1999, p. 273; Duller, 2008, p. 226):

$$F = \frac{12}{nk(k+1)} \sum_{i=1}^{k} \left(R_i - \frac{n(k+1)}{2} \right)^2$$

$$= \left(\frac{12}{nk(k+1)} \sum_{i=1}^{k} R_i^2 \right) - 3n(k+1) \, .$$

In the presence of ties, the denominator of F is corrected as follows (Hollander and Wolfe, 1999, p. 274):

$$F^* = \frac{12 \sum\limits_{i=1}^{k} R_i^2 - 3n^2 k(k+1)^2}{nk(k+1) + (1/(k-1)) \sum\limits_{i=1}^{n} (k - \sum\limits_{j=1}^{g_i} t_{i,j}^3)} \, .$$

Here, g_i denotes the number of tied groups in block i and $t_{i,j}$ the number of observations in the j-th tied group of the i-th block. If there are no ties within the blocks, we have $t_{i,j} = 1$ for all i and j, and therefore $g_i = k$, so that $F = F^*$ holds.

For large sample sizes, the test statistics F and F^*, respectively, are, under the null hypothesis, approximately χ^2 distributed with $k - 1$ degrees of freedom. Hence, an asymptotic test is possible. Moreover, an exact or approximate permutation test is possible with the test statistic F, or F^*. In total there are $(k!)^n$ permutations as the observations can be permuted within blocks only.

For illustration we again consider the data set presented in Table 9.1. Now we assume that all four treatments were applied to each person. Furthermore, we assume that each column presents the paired values of one person. Thus, we have $k = 4$ and $n = 7$. The four rank sums are $R_1 = 7.5$, $R_2 = 14.5$, $R_3 = 20.5$, and $R_4 = 27.5$ and, hence, $F = 18.69$. However, due to the ties, the statistic F^* is computed: $F^* = 19.82$. Based on the χ^2 distribution with df $= 3$, an asymptotic p-value of 0.0002 results. For the exact permutation test, the p-value is ≤ 0.0001.

For binary data the test statistic of the Friedman test can be simplified; the resulting test is called Cochran's Q test. However, the Q test is just a special case of the Friedman test; the implementation in SAS (see below) is identical (Duller, 2008, pp. 230–232).

When the grouping variable is ordinal, a trend test might be appropriate. As in the case of the Friedman test, the ranks are computed within the blocks and the k rank sums R_i are needed. Then, the test statistic of Page's trend test is (Hollander and Wolfe, 1999, p. 285)

$$L = \sum_{i=1}^{k} iR_i .$$

When it is expected under the alternative that the observations increase with an increasing group index, large values of L give evidence against the null hypothesis.

With the Page statistic L a permutation test can be performed—based on all $(k!)^n$ possible permutations or based on a random sample of permutations. For an asymptotic test, the statistic L must be standardized with

$$E_0(L) = \frac{nk(k+1)^2}{4} \quad \text{and} \quad \text{Var}_0(L) = \frac{nk^2(k+1)(k^2-1)}{144}$$

(Hollander and Wolfe, 1999, p. 285). The statistic $(L - E_0(L))/\sqrt{\text{Var}_0(L)}$ is asymptotically standard normal under the null hypothesis.

Let us consider the example of the Ames assay (Table 9.2) for illustration. Again we assume that the columns of Table 9.2 are separate blocks. The different doses are regarded as ordinal, as in Section 9.2. In this example, we have $k = 6$ and $n = 3$. The rank sums R_i are 6, 3, 11, 14, 18, and 11, so that $L = 257$ results. Moreover, we can calculate $E_0(L) = 220.5$, $\text{Var}_0(L) = 183.75$, and the standardized test statistic $(257-220.5)/\sqrt{183.75} = 2.69$. The resulting (one-sided) asymptotic p-value is 0.0035. The p-value of the permutation test is a bit smaller: 0.0023.

If there are ties within blocks, mean ranks can be used. Such ties reduce the variance of L. In the example there are no ties. If one used the formula $nk^2(k + 1)(k^2 - 1)/144$ for the variance $\text{Var}_0(L)$ and for the standardization of the test statistic, one would decrease both the test statistic and therefore the probability of rejecting the null hypothesis. As a consequence, the p-value based on the uncorrected variance would be too large, hence waiving the correction would not violate the significance level. Further details about the distribution of L for samples with and without ties can be found in Van de Wiel and Di Bucchianico (2001).

9.3.1   Implementation in SAS

The asymptotic Friedman test can be carried out with the SAS procedure FREQ. Within the TABLES statement, the variables must be listed in the following order: block (stratification) variable, group (treatment), observations. Moreover, the options CMH2 and SCORES=RANK are necessary.

```
DATA tab9_1;
INPUT person group point @@;
CARDS;
1 1 9 2 1 11 3 1 7 4 1 8 5 1 9 6 1 9 7 1 10
1 2 11 2 2 12 3 2 8 4 2 9 5 2 10 6 2 11 7 2 10
1 3 12 2 3 13 3 3 11 4 3 9 5 3 10 6 3 12 7 3 12
1 4 15 2 4 17 3 4 15 4 4 10 5 4 16 6 4 14 7 4 12
;
RUN;

PROC FREQ;
 TABLES person*group*point / CMH2 SCORES=RANK;
run;
```

In the output the Friedman statistic is given in the line "Row Mean Scores Differ." The value of the test statistic is displayed in the column "Value" and the asymptotic p-value in the column "Prob."

```
Summary Statistics for group by point
Controlling for person
```

```
Cochran-Mantel-Haenszel Statistics (Based on Rank Scores)

Statistic   Alternative Hypothesis    DF     Value       Prob

    1       Nonzero Correlation        1    19.8000     <.0001
    2       Row Mean Scores Differ     3    19.8182     0.0002

Total Sample Size = 28
```

An SAS program that can perform the Friedman test as an approximate permutation test can be found at www.webpages.uidaho.edu/~chrisw/stat514/ch4rcbcandy2.sas. This program can also be used to carry out Page's trend test.

Chapter 10

Independence and Correlation

10.1 The χ^2 Test

When observing two variables X and Y, often the question arises whether or not X and Y are independent. For example, is gender associated with voting behavior? In order to tackle such a question, the data can be summarized within a contingency table. If X and/or Y are continuous, one must create categories. Then, Pearson's χ^2 test, already introduced for a 2xk table in Chapter 5, can be applied. The null hypothesis states that the two variables X and Y are independent. Under the alternative, there is an association.

In general we have an mxk table; the notation is defined in Table 10.1. The test statistic of the χ^2 test is

$$X^2 = \sum_{i=1}^{m} \sum_{j=1}^{k} \frac{\left(x_{ij} - \frac{n_{i.}\,n_{.j}}{N}\right)^2}{\frac{n_{i.}\,n_{.j}}{N}} \, .$$

This test statistic compares the observed frequencies x_{ij} with the corresponding values expected under independence. Under the null hypothesis, the probability for the category i of X can be estimated by $n_{i.}/N$. Analogously, the probability for the category j of Y can be estimated by $n_{.j}/N$. Then, because of the independence, the probability for the combination (i, j) is the product of the two probabilities. To compute the expected frequency for the combination (i, j), this product must be multiplied with the total sample size N. The resulting expected count is $n_{i.}\,n_{.j}/N$, which is used above in the statistic X^2.

The test statistic X^2 is, under the null hypothesis, asymptotically χ^2 distributed with $(m-1)(k-1)$ degrees of freedom. Hence, in an asymptotic test the independence can be rejected if X^2 is at least as large as the $(1-\alpha)$ quantile of this χ^2 distribution. Alternatively, an exact test can be performed. In an exact conditional test the marginal totals $n_{i.}$ and $n_{.j}$ are fixed (see chapter 5).

Let us consider extra-pair paternities in birds as an example, as in Section 7.4. Saino et al. (1999) determined sex and paternity in 74 young barn swallows (*Hirundo rustica*). The results can be found in Table 10.2. The aim is to test

TABLE 10.1: Notation for an $m \times k$ contingency table

	Category of Variable Y				
Variable X	1	2	...	k	Sum
Category 1	x_{11}	x_{12}	...	x_{1k}	$n_{1.}$
Category 2	x_{21}	x_{22}	...	x_{2k}	$n_{2.}$
...					
Category m	x_{m1}	x_{m2}	...	x_{mk}	$n_{m.}$
Sum	$n_{.1}$	$n_{.2}$	...	$n_{.k}$	N

TABLE 10.2: A 2×2 table: Sex and paternity of young barn swallows

	Sex		
Paternity	Male	Female	Sum
Extra-pair	26	23	49
Not extra-pair	15	10	25
Sum	41	33	74

Source Saina et al. 1999.

the relationship between sex and the variable "paternity" for independence. The test statistic X^2 takes on the value 0.323. Because all expected frequencies are larger than 5, one can apply the asymptotic test (cf. Chapter 5). With one degree of freedom, the resulting p-value is 0.5701. The p-value of the exact permutation test is 0.6271.

This example is only used to illustrate the χ^2 test. It should be noted that the young birds are grouped within broods, and the resulting clustered data structure might be taken into account to avoid invalid conclusions (Krackow and Tkadlec, 2001).

In SAS the χ^2 test can be carried out using the procedure FREQ (see Chapter 5). If the null hypothesis can be rejected, one can conclude that there is a dependence. However, it should be noted that a significant test does not provide evidence for a causal relationship. In Chapter 5 the χ^2 test was applied in order to test for the homogeneity of two distributions. That test, including the test statistic, distribution, and rejection region, is identical to the χ^2 test for independence described here; this equality also holds for more than two groups.

10.2 The Likelihood-Ratio Test

The likelihood-ratio test G is an alternative to the χ^2 test. Its test statistic G is defined as (see, e.g., Zar, 2010, p. 509)

$$G = 2 \sum_{i=1}^{m} \sum_{j=1}^{k} x_{ij} \ln \left(\frac{x_{ij}}{e_{ij}} \right) ,$$

where e_{ij} are the expected frequencies under the null hypothesis. Here we have $e_{ij} = (n_{i.} n_{.j})/N$.

As with the χ^2 test, one can perform an exact or an asymptotic test with the statistic G. As with X^2, the statistic G is, under the null hypothesis, asymptotically χ^2 distributed with $(m-1)(k-1)$ degrees of freedom.

With the example data of Table 10.2 we obtain $G = 0.3240$, and 0.5692 results as the asymptotic p-value. The p-value of the exact permutation test is 0.6271.

When the exact χ^2 test is requested with the statement EXACT CHISQ of the SAS procedure FREQ, the exact likelihood-ratio test G is also displayed in the output:

```
        Pearson Chi-Square Test


Chi-Square                      0.3226
DF                                   1
Asymptotic Pr >  ChiSq          0.5701
Exact        Pr >= ChiSq        0.6271



        Likelihood Ratio Chi-Square Test


Chi-Square                      0.3240
DF                                   1
Asymptotic Pr >  ChiSq          0.5692
Exact        Pr >= ChiSq        0.6271
```

When (at least) one cell of the contingency table has no observation, that is, $x_{ij} = 0$, the statistic G cannot be computed because the logarithm of 0 is not defined. If all x_{ij} are positive, both tests, the χ^2 test and the likelihood-ratio test, are possible and the question arises as to which test is preferable. The answers in the literature are not consistent (see, e.g., Zar, 2010, p. 510). Recently, Ruxton and Neuhäuser (2010b) suggested χ^2 for simple tests of association in a two-dimensional table such as that presented in Table 10.1.

Possible continuity corrections are not discussed here because exact tests are preferable in the case of small sample sizes.

10.3   Correlation Coefficients

The correlation is a measure of the degree of relationship between two variables X and Y. When the raw values are denoted by x_i and y_i, $i = 1, \ldots, n$, and their means by $\bar{x}$ and $\bar{y}$, the correlation coefficient r is defined as

$$r = \frac{\sum\limits_{i=1}^{n} (x_i - \bar{x})(y_i - \bar{y})}{\sqrt{\sum\limits_{i=1}^{n} (x_i - \bar{x})^2 \sum\limits_{i=1}^{n} (y_i - \bar{y})^2}}.$$

This correlation coefficient r is also called Pearson's product-moment correlation coefficient; its range is the interval from -1 to 1. If there is no correlation between two variables, the correlation coefficient r is zero. Positive values of r indicate accordance. For example, height is positively correlated with weight because weight increases with height—at least on average. The stronger the accordance between two variables, the larger r is; that is, the closer r is to 1. In the case of a negative correlation, one variable decreases when the other one increases, and $r < 0$. An example of a negative correlation is the association between scored goals and goals against. Good soccer teams usually score many goals, but collect only a few goals against, and vice versa for bad teams.

Pearson's product-moment correlation coefficient quantifies a linear relationship only. When there is a nonlinear relationship between two variables, r can be small or even zero.

Correlation does not imply causation. Even a very large correlation cannot be used to infer a causal relationship between the two variables. There can be a causal relation in the case of a high correlation, but a high correlation is not sufficient for establishing a causal relation. An example of a spurious correlation is that between the density of head hair and income in men. The correlation is caused by a third variable, the age, which influences both hair and income (Krämer, 1992, p. 146).

Another example is the correlation between latitude and species richness. A higher diversity is found at low latitudes, especially in the tropics. There are different "third" variables in this example: "it is obvious that latitude ... cannot itself have a *direct* effect on the number of species. However, many variables such as climate and land or ocean area are correlated with latitude and may provide an explanation for tropical diversity" (Sherratt and Wilkinson, 2009, p. 100).

The effect of a third variable can be removed within the concept of partial correlation. The partial correlation is the correlation between X and Y after partialing out the influence of the third variable U (Hartung et al., 2009, p. 561). The partial correlation $r_{XY|U}$ can be computed based on the usual correlation coefficients:

$$r_{XY|U} = \frac{r_{XY} - r_{XU}r_{YU}}{\sqrt{(1 - r_{XU}^2)(1 - r_{YU}^2)}} \,,$$

where the index shows the corresponding variables, that is, r_{XU} denotes the correlation coefficient between X and U.

When the aim is to test whether a correlation exists, one can test the null hypothesis that the correlation is zero. A permutation test can be used to test such a null hypothesis. Obviously, the data are paired. However, it makes no sense to permute the two values of a pair, as done when performing the Wilcoxon signed rank test. The two values of a pair are observations of different variables. Now, one variable, say the x_i values, can remain unchanged. And the values of the other variable are permuted, and every y_i is assigned to every x_i once. Because there is no correlation under the null hypothesis, each y value could be observed with equal probability, together with any x value. Hence, there are $n!$ different permutations.

For every permutation, the test statistic r must be calculated. In the case of a two-sided test, the p-value is the proportion of permutations for which the absolute value of r is at least as large as for the actually observed data.

The correlation coefficient r can also be computed as

$$r = \frac{\sum\limits_{i=1}^{n} x_i y_i \;-\; n\bar{x}\bar{y}}{\sqrt{\left(\sum\limits_{i=1}^{n} x_i^2 - n\bar{x}^2\right)\left(\sum\limits_{i=1}^{n} y_i^2 - n\bar{y}^2\right)}} .$$

Except $\sum\limits_{i=1}^{n} x_i y_i$, all terms in this formula are constant for all permutations. Thus, this sum $\sum\limits_{i=1}^{n} x_i y_i$ can be used as the test statistic of the permutation test.

The test statistic L of Page's trend test presented in Section 9.3 is a special case of this sum $\sum x_i y_i$. In order to see this, the x_i take on the values 1, 2, ..., k, and the y_i corresponds to the rank sums R_i (Higgins, 2004, p. 151).

As an example, we consider data about the carotin content of grass in relationship to the duration of storage (Steger and Püschel, 1960; Rasch and Verdooren, 2004, p. 16) (see Table 10.3). For these data, $n = 5$ and

TABLE 10.3: Carotin content of grass in relationship to the duration of storage

Duration of Storage (in days)	Carotin Content (in mg/100g)
1	31.25
60	28.71
124	23.67
223	18.13
303	15.53

Source Steger and Fuschei 1960; Rascn and Verdooren 2004.

TABLE 10.4: The exact permutation null distribution of r for the example of Steger and Püschel (1960)

Possible Value of r	Probability (= proportion within the 120 permutations)
−0.9923	1/120
−0.9464	1/120
−0.9286	1/120
−0.8935	1/120
−0.8827	1/120
...	...
0.8799	1/120
0.8806	1/120
0.9269	1/120
0.9422	1/120
0.9891	1/120

$r = -0.9923$. Thus, there is a strong negative correlation. For the permutation test, $n! = 5! = 120$ permutations must be considered. The resulting permutation null distribution of r is displayed in Table 10.4. The observed data set corresponds to an extreme permutation, a value smaller than -0.9923 does not occur. Furthermore, the absolute value $|r|$ is smaller for all other permutations. The permutation null distribution is not symmetric; there is no permutation with a positive correlation of 0.9923. The two-sided p-value therefore is $1/120 = 0.0083$ and is equal to the probability $P_0(|r| \geq 0.9923)$. The null hypothesis, which states that the correlation is zero, can be rejected at the level $\alpha = 0.05$.

For an asymptotic test, one can use the test statistic $\sqrt{(n-2)/(1-r^2)}\, r$. If the variables X and Y are bivariate normal, this test statistic has, under the null hypothesis, a t distribution with $n - 2$ degrees of freedom (see, e.g.,

Higgins, 2004, p. 146). In the example we have -13.85 as the value of the test statistic, and a two-sided p-value of 0.0008 results.

When the usual correlation coefficient r is computed with the ranks, the resulting coefficient is called Spearman's rank correlation coefficient. This coefficient can be used not only for quantitative, but also for ordinal data. In order to compute the rank correlation coefficient r_S, one must determine the ranks for the x_i and separately for the y_i. Then, the ranks can be entered into the formula for r given above in order to get Spearman's rank correlation. If ties occur within the x_i, or within the y_i, respectively, one can use mean ranks. When there are no ties, a simpler formula can be applied; to be precise,

$$r_S = 1 - \frac{6 \sum_{i=1}^{n} d_i^2}{(n-1)n(n+1)},$$

where d_i is the difference rank of x_i – rank of y_i (see, e.g., Büning and Trenkler, 1994, p. 233).

As with Pearson's correlation coefficient r, Spearman's rank correlation coefficient r_S also has a range from -1 to 1. The rank correlation coefficient is 1 or -1 not only in case of a strong linear relationship, but also when there is a monotone accordance. To be precise, r_S is 1 if the two rankings are identical, and r_S is -1 if one ranking is the reverse of the other. The latter is the case for the example data of Table 10.3; thus $r_S = -1$.

The permutation test described above can also be performed with the test statistic r_S. In the example, the observed data set is an extreme permutation also with regard to the test statistic r_S. However, now the permutation null distribution of r_S is symmetric (Hollander and Wolfe, 1999, p. 400), and there is also a permutation with $r_S = 1$ (see Table 10.5), The two-sided p-value therefore is $P_0(|r_S| \geq 1) = 2/120 = 0.0167$. Again, the null hypothesis can be rejected for $\alpha = 0.05$.

In the case of a large sample, the permutation test can also be performed as an approximate test based on a simple random sample of permutations. Moreover, for an asymptotic test, one can use that $\sqrt{n-1}\, r_S$ is, under the null hypothesis, asymptotically (for $n \to \infty$) standard normal (see, e.g., Hollander and Wolfe, 1999, p. 395).

10.3.1   Implementation in SAS

In SAS the correlation coefficients r and r_S can be computed using the procedure CORR. However, in order to carry out the permutation tests, the procedure FREQ is required. The exact test with the statistic r can be invoked with the statement EXACT PCORR:

TABLE 10.5: The exact permutation null distribution of r_S for the example of Steger and Püschel (1960)

Possible Value of r_S	Probability (= proportion within the 120 permutations)
−1	1/120
−0.9	4/120
−0.8	3/120
−0.7	6/120
−0.6	7/120
−0.5	6/120
−0.4	4/120
−0.3	10/120
−0.2	6/120
−0.1	10/120
0	6/120
0.1	10/120
0.2	6/120
0.3	10/120
0.4	4/120
0.5	6/120
0.6	7/120
0.7	6/120
0.8	3/120
0.9	4/120
1	1/120

```
PROC FREQ;
  TABLES duration*carotin;
  EXACT PCORR;
RUN;
```

The output, displayed here in parts, gives the correlation coefficient r as well as the result of the exact permutation test:

```
Pearson Correlation Coefficient

Correlation (r)         -0.9923
ASE                      0.0034
95% Lower Conf Limit    -0.9990
95% Upper Conf Limit    -0.9856

  Test of H0: Correlation = 0
```

```
Exact Test
One-sided Pr <=  r        0.0083
Two-sided Pr >= |r|       0.0083
```

In order to perform the permutation test with the statistic r_S, the statement EXACT SCORR rather than EXACT PCORR is required.

Chapter 11

Stratified Studies and Combination of p-Values

11.1 The Van Elteren Test

Clinical trials are often multicentric, that is, patients are recruited in several centers (medical practices or hospitals). In so doing, a larger sample size is possible, among other things. A disadvantage, however, is the possibility of differences between the centers. Therefore, the randomization of patients is usually done separately for every center in order to have in each center a similar allocation of patients to the different treatments. A consequence for the statistical analysis is that the stratification by center must be considered (EMEA, 2003).

In an analysis of covariance, the center could be included as an additional factor. A very general alternative is to calculate the test statistic separately for each value of the factor, that is, each center, and to compute the sum of the single test statistics (Mehta et al., 1992). Such a stratified test is, for instance, possible with the Wilcoxon rank sum test. The resulting stratified Wilcoxon test is also called the Van Elteren test (Van Elteren, 1960).

When applying the Van Elteren test, the Wilcoxon rank sums are determined separately for each center (or in general for each stratum in a stratified design) and added up. Let k be the number of centers. The distribution function for center i, $i = 1, \ldots, k$, and treatment (group) j, $j = 1$, 2, is denoted by $F[(x - \theta_{ij})/\theta_i]$, where the shift parameter θ_{ij} depends on center and treatment. In contrast, the scale parameter θ_i depends on center only. Hence, a difference in location is possible between the treatments. The extent of such a shift in location can differ between centers. Moreover, in this model, differences in variability are possible between centers. Within a center, the variances are identical for both treatments.

The null hypothesis is $\theta_{i1} = \theta_{i2}$ for all i. Thus, there is no center with a difference in location between the two treatments. A possible one-sided alternative hypothesis states $\theta_{i1} \leq \theta_{i2}$ for all i and $\theta_{i1} < \theta_{i2}$ for at least one center.

As mentioned above, the Wilcoxon rank sum is computed separately for each center; the resulting rank sums are denoted by $W_i, \ldots, W_k$. The sum of these Wilcoxon rank sums over all centers is $T = \sum_{i=1}^{k} W_i$. The center-specific rank sums W_i can be weighted. Van Elteren (1960) suggested the weighting scheme $\sum_{i=1}^{k} W_i/(n_{1i} + n_{2i} + 1)$, where n_{1i} and n_{2i} denote the sample sizes of the two groups in center i. As Bradley (1968, pp. 115–116) did, we will use the unweighted sum T.

Let $\mathrm{E}_0(W_i)$ and $\mathrm{Var}_0(W_i)$ be the expected values and variances of W_i under the null hypothesis, they can be calculated according to the formulas presented in Section 2.2. Because the centers are independent, the expected value and variance of T are sums:

$$\mathrm{E}_0(T) = \sum_{i=1}^{k} \mathrm{E}_0(W_i) \quad \text{and} \quad \mathrm{Var}_0(T) = \sum_{i=1}^{k} \mathrm{Var}_0(W_i).$$

The standardized statistic $(T - \mathrm{E}_0(T))/\sqrt{\mathrm{Var}_0(T)}$ is asymptotically standard normal and can be used for an asymptotic test. A permutation test can be performed with the statistic T. However, the number of permutations is very large in a stratified design. The reference set for the permutation test is the Cartesian product of the sets for the single centers. Hence, in total there are $\prod_{i=1}^{k} \binom{N_i}{n_{1i}}$ permutations, where n_{1i} denotes the sample size in group 1 and center i, and N_i denotes the total sample size in center i. Due to the huge number of permutations, a permutation test in a stratified design is usually possible only with the help of efficient algorithms (see Mehta et al., 1992).

11.1.1   Example

As an example we consider a clinical trial; a new drug for the treatment of psoriasis was compared to a placebo in nine centers. The clinical endpoint is an ordered categorical variable with possible scores of 1 to 5. Score 1 is the worst result and score 5 the best. The raw data were given by Boos and Brownie (1992) and are reproduced here in Table 11.1. Table 11.1 also displays the Wilcoxon rank sums W_i; these are calculated by summing up the ranks of the new treatment, that is, the "verum."

In this example the sum $T = \sum_{i=1}^{9} W_i$ is 643. With $\mathrm{E}_0(T) = 556$ and $\mathrm{Var}_0(T) = 701.98$, the standardized statistic is $(T - \mathrm{E}_0(T))/\sqrt{\mathrm{Var}_0(T)} = 3.2837$. The resulting asymptotic p-value is $1 - \Phi(3.2837) = 0.0005$ for the one-sided test with the alternative that the scores are larger under the new treatment.

TABLE 11.1: A multicenter placebo-controlled clinical trial in patients with psoriasis:

Center	Group	1	2	3	4	5	W_i	$E_0(W_i)$	$\sqrt{\text{Var}_0(W_i)}$
		\multicolumn{5}{c}{—— Score ——}	Rank Sum		
1	Placebo	1	3	4	2	0			
	Verum	0	1	5	4	0	123.5	105	12.38
2	Placebo	0	2	2	1	0			
	Verum	0	1	2	1	0	21.5	20	3.82
3	Placebo	0	3	2	3	0			
	Verum	0	0	0	5	3	92.5	68	8.85
4	Placebo	0	1	5	3	0			
	Verum	0	2	3	3	0	70	72	9.58
5	Placebo	0	3	3	5	0			
	Verum	0	3	4	3	1	125	126.5	14.46
6	Placebo	0	1	6	0	0			
	Verum	0	1	4	1	2	73.5	64	7.24
7	Placebo	0	1	2	0	0			
	Verum	0	2	1	0	0	9	10.5	2.01
8	Placebo	0	4	4	0	0			
	Verum	0	0	2	6	0	96	68	8.94
9	Placebo	0	2	4	0	0			
	Verum	0	0	1	3	0	32	22	4.32

Source Boos and Brownie 1992.

Dmitrienko et al. (2005, pp. 16–19) describe how the Van Elteren test can be carried out with the SAS procedure FREQ. Further details about the Van Elteren test, as well as modified methods that have some advantages, are presented by Thangavelu and Brunner (2007).

11.2 Combination Tests

The Van Elteren test connects test statistics. Combination tests, in contrast, combine p-values of independent tests. Let us assume that the k null hypotheses H_{0i} are tested versus the alternatives H_{1i}, $i = 1, \ldots, k$. The k p-values are denoted by $p_1, \ldots, p_k$. These p-values are based on k independent tests; for example, there might be k different studies. A combination test now examines the global null hypothesis that all k null hypotheses H_{0i} are true, versus the global alternative that at least one alternative H_{1i} is true. It is possible that all single null hypotheses are identical.

A simple combination test is based on the minimum of the k p-values. This method, according to Tippett (1931), rejects the global null hypothesis at level α if

$$\min(p_1, \ldots, p_k) \leq 1 - (1 - \alpha)^{1/k} \, .$$

Other combination tests utilize the fact that a p-value is, under the null hypothesis, uniformly distributed on the interval $[0, 1]$ if the corresponding test statistic is continuous. This result holds regardless of the specific form of the test statistic, the underlying test problem, and the distribution of the raw data (Hartung et al., 2008, p. 25).

The inverse normal method uses the fact that $z(p_i) = \Phi^{-1}(p_i)$ is, under the null hypothesis, standard normal. Because the tests and therefore the p-values are independent, the sum $\sum_{i=1}^{k} z(p_i)/\sqrt{k}$ is also standard normal. Thus, the inverse normal method rejects the global null hypothesis at level α if

$$\frac{1}{\sqrt{k}} \sum_{i=1}^{k} \Phi^{-1}(p_i) \leq -z_\alpha \, .$$

Here, a small value of the sum gives evidence against the null hypothesis. The reason is that $z(p_i)$ is small for small p-values.

Let us consider the case that there are no ties and that there is no difference in sample size between two groups to be compared in all strata (centers). Then, $(T - E_0(T))/\sqrt{Var_0(T)}$, the standardized test statistic of the unweighted Van Elteren test, is equal to the sum of the standardized statistics divided by $\sqrt{k}$; that is,

$$\frac{1}{\sqrt{k}} \sum_{i=1}^{k} \frac{(W_i - E_0(W_i))}{\sqrt{Var_0(W_i)}} \, .$$

In this case, the unweighted Van Elteren test matches the inverse normal method; both tests give identical results (Neuhäuser and Senske, 2009).

Fisher's combination test is a further method that was mentioned in Van Elteren's paper as an alternative (van Elteren, 1960). Because p_i has a uniform distribution on $[0, 1]$, $-2 \ln p_i$ is χ^2 distributed with two degrees of freedom. Because of the independence, it follows that $-2 \sum_{i=1}^{k} \ln p_i$ has a χ^2 distribution with df $= 2k$. Hence, the global null hypothesis can be rejected at level α if

$$-2 \sum_{i=1}^{k} \ln p_i \geq \chi^2_{2k, 1-\alpha} \, ,$$

where $\chi^2_{2k, 1-\alpha}$ is the $(1 - \alpha)$ quantile of the χ^2 distribution with $2k$ degrees of freedom. This decision rule can also be formulated based on the product of the p-values: The global null hypothesis is rejected at level α if

$$\prod_{i=1}^{k} p_i \leq \exp\left(-\frac{\chi^2_{2k,1-\alpha}}{2}\right) .$$

For $k = 2$ and $\alpha = 0.05$, we have $\chi^2_{4,0.95} = 9.4877$; hence, the product $p_1 p_2$ must not be larger than 0.0087 for a significance. For example, this is fulfilled for $p_1 = p_2 = 0.09$: $p_1 p_2 = 0.0081 < 0.0087$. Therefore, two tests, both nonsignificant at level 5% when considered separately, can give a significance in the combination test. This is no contradiction. One cannot argue that the nonsignificant result of the first test is reproduced and therefore confirmed by the nonsignificant result of the second test. A failed significance is never a secured result; it was just not possible to reject the null hypothesis. And the two p-values of 0.09 are not large; they indicate that there might be a deviation from the null hypothesis. However, this deviation was not strong enough for a significance at the 5% level. But the evidence is stronger when both tests are considered together. Then, the combination test is significant in this case.

There are further methods for the combination of p-values; see Hartung et al. (2008, Chapter 3). However, no combination test can be recommended over all others in general. According to Hartung et al. (2008, p. 28), the inverse normal method is widely used in the social sciences. When applying this method, two p-values with $p_i = 1 - p_j$ neutralize because $\Phi^{-1}(p_i) + \Phi^{-1}(1 - p_i) = 0$. In contrast, when applying Fisher's combination test individual small p-values can be sufficient for a significance, even when some other p-values are large. For instance, in case of $k = 2$, the two p-values $p_1 = 0.005$ and $p_2 = 0.995$ are together significant in Fisher's test as $p_1 p_2 = 0.0050 < 0.0087$—although $p_1 = 1 - p_2$ holds. Consequently, Fisher's combination test has a larger sensitivity for data that dismiss a null hypothesis compared to data that support it (Rice, 1990). This can be an advantage. In the situation that there is no specific alternative, Fisher's combination test can be recommended (see Hartung et al., 2008, and references given there). For the distribution of Fisher's combination statistic when the tests are dependent, the reader is referred to Yang (2010).

11.2.1   Example

Let us consider the four p-values $p_1 = 0.0042$, $p_2 = 0.0150$, $p_3 = 0.5740$, and $p_4 = 0.1127$ as an example. These are the p-values of Wilcoxon rank sum tests in a multicenter clinical trial designed to compare sulpiride and placebo (Rüther et al., 1999; Neuhäuser and Senske, 2009).

For the inverse normal method, the quantiles $z(p_i) = \Phi^{-1}(p_i)$ must be computed: $z(p_1) = \Phi^{-1}(0.0042) = -2.6356$, $z(p_2) = -2.1701$, $z(p_3) = 0.1866$, and $z(p_4) = -1.2123$. Hence, the sum is $\sum_{i1}^{k} z(p_i)/\sqrt{k} = -2.9157$, and a p-value of $\Phi(-2.9157) = 0.0018$ results for the inverse normal method.

For Fisher's combination test we must calculate $-2\sum_{i=1}^{k} \ln p_i = 24.8211$, the corresponding p-value based on the χ^2 distribution with df $= 8$ is 0.0017. It should be noted that the unweighted Van Elteren test gives the p-value 0.0023 in this example (Neuhäuser and Senske, 2009).

In SAS, Fisher's combination test can be performed with the procedure PSMOOTH. In the procedure, calling the option FISHER is required. Moreover, a VAR statement is necessary as for the procedure NPAR1WAY (see Section 2.1). This VAR statement identifies the variable that contains the p-values to be combined. In order to carry out the inverse normal method, one can use the function Φ^{-1} called PROBIT in SAS.

11.3   A Combination Test for Discrete Test Statistics

Fisher's combination test can also be applied to independent p-values of discrete test statistics. In that case, the combination test is often conservative (see Mielke et al., 2004, and references given there). As an alternative, Mielke et al. (2004) recommended a method described below with an example.

For a multinomial distribution with three possible outcomes, the null hypothesis $\pi_1 = 0.1$, $\pi_2 = 0.35$, and $\pi_3 = 0.55$ should be tested, where π_i denotes the probability for category i. In a first study with two observations, the first two categories were observed once. The vector of observed frequencies therefore is $(1, 1, 0)$. The frequencies expected under the null hypothesis are $(2\pi_1, 2\pi_2, 2\pi_3) = (0.2, 0.7, 1.1)$. The resulting χ^2 test statistic is

$$X_1^2 = \frac{(1-0.2)^2}{0.2} + \frac{(1-0.7)^2}{0.7} + \frac{(0-1.1)^2}{1.1} = 4.429 \,.$$

In a second study, again with two observations, the first category was observed twice. Hence, the vector of observed frequencies is $(2, 0, 0)$. The expected frequencies are unchanged and the value of the test statistic X_2^2 is 18.

In addition to the two observed vectors, four other vectors are possible. All six vectors of observed frequencies are listed in Table 11.2, together with the corresponding values of the test statistic and the probabilities under the null hypothesis. In the second study, an extreme value of the test statistic was observed. The p-value of the exact χ^2 test for the second study therefore

TABLE 11.2: Possible vectors of observed frequencies, probabilities under the null hypothesis, and values of the χ^2 test statistic for a multinomial distribution with three outcomes

Number	Vector	Probability	χ^2 Statistic
1	$(0,0,2)$	0.3025	1.636
2	$(0,1,1)$	0.3850	0.338
3	$(0,2,0)$	0.1225	3.714
4	$(1,0,1)$	0.1100	3.909
5	$(1,1,0)$	0.0700	4.429
6	$(2,0,0)$	0.0100	18.00

is equal to the probability 0.0100 for the actually observed vector $(2,0,0)$ (number 6). In the first study, vector 5, that is $(1,1,0)$, was observed. For this vector the test statistic attains its second-largest value. The p-value of the exact χ^2 test for study 1 therefore is $0.0100 + 0.0700 = 0.0800$.

With the usual combination test of Fisher we have $-2(\ln(0.01) + \ln(0.08)) = 14.26$. Based on the χ^2 distribution with 4 degrees of freedom, the p-value is 0.0065.

For the modification proposed by Mielke et al. (2004) for discrete test statistics, the sum of the χ^2 test statistics over the strata must be computed: $4.429 + 18.00 = 22.429$. In order to determine the p-value of the combination test, one must identify all combinations that give a sum at least as large as the actually observed one. In the example, that is, the observed combination of the vectors 5 and 6, and the combinations 6 and 5 as well as 6 and 6. Because of the independence of the two studies, the probabilities for the combinations can be computed as products. To be precise, $0.07 \cdot 0.01 = 0.0007$ is the probability for the combination 5 and 6. The combination 6 and 5 has the same probability, whereas the combination 6 and 6 has the probability $0.01^2 = 0.0001$. The sum of these three probabilities, that is 0.0015, is the p-value of the modified combination test.

This analogue to Fisher's combination test proposed for discrete test statistics (Mielke et al., 2004) is merely a stratified test where the test statistic is, as usual, the sum over the strata. We will illustrate this principle using two further examples. The first one is an application of χ^2 statistics with two strata, as was presented by Neuhäuser (2003d).

The sex of the oldest nestling in broods of the kakapo (*Strigops habroptilus*) was investigated. The null hypothesis states a balanced sex ratio, that is, both sexes occur with identical probabilities. There are two strata that differ by the feeding of the female. In the first stratum (fed females) there are five observations, and two in the second stratum (unfed females). In both

TABLE 11.3: Observed and all other possible tables, expected frequencies under the null hypothesis of a balanced sex ratio, values of the χ^2 test statistic, and probabilities under the null hypothesis

Table (number)	Expected Frequencies		Value of χ^2 Statistic	Proba-bility
Stratum 1				
5 0 (1)	**2.5**	**2.5**	**5.0**	**0.03125**
4 1 (2)	2.5	2.5	1.8	0.15625
3 2 (3)	2.5	2.5	0.2	0.3125
2 3 (4)	2.5	2.5	0.2	0.3125
1 4 (5)	2.5	2.5	1.8	0.15625
0 5 (6)	2.5	2.5	5.0	0.03125
Stratum 2				
2 0 (7)	1	1	2.0	0.25
1 1 (8)	1	1	0.0	0.5
0 2 (9)	**1**	**1**	**2.0**	**0.25**
Observed value of the stratified test statistic			**5.0 + 2.0 = 7.0**	
Tables with an equal or greater value of the stratified test statistic in comparison to the observed combination (1, 9) (numbers acc. to row numbers given above)			(1, 7) (1, 9) (6, 7) (6, 9)	0.0078125[a] 0.0078125[a] 0.0078125[a] 0.0078125[a]

[a] Product of the probabilities of the two respective tables.

strata only young of one sex were observed; hence, extreme tables occurred in both strata. Table 11.3 displays all possible tables with fixed sample sizes as well as the corresponding values of the test statistic and the probabilities. The *p*-value of the stratified test is computed as the sum of the probabilities of those combinations whose stratified test statistic is at least as large as the one of the actually observed combination. Hence, four combinations with a stratified test statistic ≥ 7 must be considered. The resulting *p*-value is $4 \cdot 0.0078125 = 0.03125$; thus, there is a significance at the 5% level.

Here, only nestlings of one sex occurred within each stratum. However, the sex differs between the strata, and thus the results are in the opposite direction. In this example, that is an expected situation: females in better condition, that is fed females, may produce more offspring of the larger sex (as in many other polygynous species, male kakapo typically weigh much more than females); see Neuhäuser (2003d) for further details. In other applications, however, it can be a problem when results in opposite directions do not neutralize, but cumulate and might get significant in a combination test. This must be taken into consideration when combining two-sided tests.

TABLE 11.4: Frequencies of animals with a tumor at different doses and in different time intervals in an animal tumorigenicity study

Time Interval (weeks)		Dose Level d_i		
		0	1	2
0-50	O^a	0	0	0
	C^b	1	3	3
51-80	O^a	0	0	0
	C^b	4	5	7
81-104	O^a	0	0	2
	C^b	10	12	15
Terminal	O^a	0	1	0
sacrifice	C^b	35	30	25

[a] O = observed tumor count.
[b] C = number of animals necropsied.

As an additional example we consider an animal tumorigenicity study; three doses were investigated within each stratum (Lin and Ali, 1994, p. 33). Here, a trend test can be applied. The study is partitioned into four time intervals (= strata). The three doses are a control group ($d_0 = 0$), a low dose ($d_1 = 1$), and a high dose ($d_2 = 2$). The observed frequencies of animals with a tumor are displayed in Table 11.4.

In order to adjust for a possible effect of the covariable time, a stratified test with the time intervals as strata is performed. The test statistic of the stratified test is the sum of the statistics over the strata. Because the sample sizes are small and the numbers of tumors are very low, an exact test is appropriate. The reference set for the permutation test again is the Cartesian product of the sets for the single strata. As Mehta et al. (1992) we consider a conditional test. The null hypothesis states that there is no difference between the doses.

As no tumor occurs in the first two intervals (see Table 11.4), these intervals contribute nothing to the test statistic and only the last two intervals must be considered (Lin and Ali, 1994; Neuhäuser and Hothorn, 1999). For these last two intervals, Table 11.5 displays the observed and all other possible tables with the same number of tumors per strata. Furthermore, the values of the (nonstandardized) Armitage trend test applied by Lin and Ali (1994) and the corresponding hypergeometric probabilities under the null hypothesis are shown.

The test statistic of the stratified test is the sum over the strata, that is, $4 + 1 = 5$. In order to compute the one-sided p-value, we consider the combinations whose value of the test statistic is at least as large as actually observed; these

TABLE 11.5: Analysis of the data of Table 11.4: Observed and all other possible tables with the same number of tumors per strata, values of the Armitage test statistic and hypergeometric probabilities under the null hypothesis

Table (Number)		T_{CA}[a]	Probability
Stratum 81 – 104 weeks			
0 0 2	**(1)**	**4**	**0.15766**
0 2 0	(2)	2	0.09910
2 0 0	(3)	0	0.06757
0 1 1	(4)	3	0.27027
1 0 1	(5)	2	0.22523
1 1 0	(6)	1	0.18018
Stratum terminal sacrifice			
0 0 1	(7)	2	0.27778
0 1 0	**(8)**	**1**	**0.33333**
1 0 0	(9)	0	0.38889

[a]Test statistic of the non-standardized Armitage test:
$T_{CA} = d_0 \cdot T_0 + d_1 \cdot T_1 + d_2 \cdot T_2$, where T_i is the frequency of animals with a tumor in the respective dose group.

are the combinations (1, 7), (1, 8), and (4, 7). With the probabilities given in Table 11.4 the one-sided p-value can be calculated as follows:

$$0.15766 \cdot 0.27778 + 0.15766 \cdot 0.33333 + 0.27027 \cdot 0.27778 = 0.17142.$$

If the data can be presented in a 2x2 table for each stratum, the Mantel-Haenszel test can be applied. Again, the test statistic is a sum over the strata. This test can be carried out as an exact or an asymptotic test. Details as well as an example are given, for instance, in Higgins (2004, Section 5.7).

Chapter 12

Nonstandard Situations and Complex Designs

When applying a permutation or bootstrap test, it is not necessary to determine the theoretical distribution of the test statistic. This enables great flexibility. For instance, test statistics can be used that are extremely difficult, or even impossible, to handle analytically. Moreover, very complex designs can be analyzed. As a consequence, permutation and bootstrap tests might be the method of choice for nonstandard situations and complex designs. According to Manly (2007, p. 341), it is one of the most important advantages of computer-intensive methods such as bootstrap and permutation tests, that they can be applied for data that do not fit into any of the usual categories. Manly (2007) presented four examples, including a study of Cushman et al. (1993) who investigated whether the size of European ants depends on the latitude (see Manly, 2007, pp. 365–369).

It was assumed that the relationship between the expected size $E(Y)$ and the latitude X can be modeled using a linear regression of the form $E(Y) = \alpha + \beta X$. In total, there are 2,341 pairs of observations. Nonstandard features of this example are that only mean sizes are available for each species, and that several species live at a range of different latitudes. Therefore, there are numerous pairs with different values of X, but identical values of Y.

With the observed data a slope estimate of $\hat{\beta} = 0.0673$ resulted. In order to assess the significance of this slope, Cushman et al. (1993) performed an approximate permutation test. The sizes were randomly allocated to the different species. The latitudes remained unchanged for all species. Within 9,999 permutations there were only six with a slope larger than or equal to 0.0673. Together with the original data there are 10,000 permutations; the corresponding one-sided p-value is $7/10,000 = 0.0007$. Therefore, the relationship is significant: the higher the latitude, the larger seems to be the mean size of ant species.

In the following Sections 12.1 and 12.2 two further nonstandard situations will be presented. Furthermore, it should be noted that there are so-called Monte Carlo tests that can be suitable especially in nonstandard situations. In a Monte Carlo test, the significance of an observed value of a test statistic is determined based on a comparison with a random sample of values of the test statistic that are simulated under an assumed model. Permutation as well as bootstrap tests can be regarded as special cases of Monte Carlo tests (Manly, 2007, p. 81).

As a further example of a Monte Carlo test, a test for a difference in variability is briefly described. As discussed in Section 3.4, Levene's transformation can be applied for such a test. Now suppose that a specific distribution can be assumed for the untransformed data (e.g., a normal distribution). Then, the distribution of the test statistic, calculated after the Levene transformation, can be determined via simulation. Further details of this example of a Monte Carlo test are given by Neuhäuser and Hothorn (2000).

Permutation tests can also be applied for complex designs. For example, it is possible to permute residuals (ter Braak, 1992). For a review of permutation tests for complex designs, we refer readers to Anderson (2001), Anderson and ter Braak (2003), and Manly (2007). Permutation testing for factorial designs, repeated measurements, and some other complex data are discussed in detail by Pesarin (2001) and Pesarin and Salmaso (2010). Moreover, Basso et al. (2009) present permutation tests for complex models such as two-way designs and unreplicated factorial designs.

12.1   A Contingency Table with Empty Cells

The aquatic warbler (*Acrocephalus paludicola*) is a bird species without a pair bond. The male does not assist in nest-building, incubating eggs, or feeding the young. The offspring within a single brood can be sired by several different males. In a DNA fingerprinting study, seventy nestlings from eighteen broods were investigated (see Table 12.1, Schulze-Hagen et al., 1993). All females were the genetic mothers of all young in their respective nests. Half of the eighteen broods descended from a single male in each case. Within the other nine broods there were two or more fathers. Multiple paternity occurred in all broods with five or more nestlings. That finding indicates that a female's reproductive success may increase with the number of sires.

In order to answer the question of whether multiple paternities raise the reproductive success of a female, one can apply a trend test that checks whether the number of nestlings increases with the number of sires. However, there is a problem: The number of young within a brood cannot be smaller than the number of sires. The cells marked with a minus sign (instead of a zero) in Table 12.1 must be empty. It is not possible to have an observation in such a cell. This restriction causes an inherent trend that can contribute to a significance without the need of any other effect (Neuhäuser et al., 2003).

Hence, a modified trend test is required. Due to the small sample sizes and

TABLE 12.1: Frequencies of multiple paternities in broods of the aquatic warbler

Number	Number of Nestlings per Brood					
of Sires	1	2	3	4	5	6
1	0	2	5	2	0	0
2	-	0	1	0	0	0
3	-	-	1	0	3	1
4	-	-	-	2	1	0

the ties, a permutation test is appropriate. However, we must not consider all tables with identical marginal totals. Without any further constraint there are 5,196 tables with marginal totals identical to those of Table 12.1. The probabilities of these tables can be computed under the null hypothesis that there is no trend, as follows:

$$P_0(x_{11}, \ldots, x_{46}) = \frac{\prod\limits_{i=1}^{4} n_i! \prod\limits_{j=1}^{6} m_j!}{N! \prod\limits_{i=1}^{4} \prod\limits_{j=1}^{6} x_{ij}!},$$

where x_{ij} is the frequency of the combination i sires and j nestlings. Moreover, the row sums are denoted by n_i and the column sums by m_j ($N = \sum n_i$), as defined in Table 5.2.

Because of the restriction that the number of nestlings within a brood cannot be smaller than the number of sires, we have $x_{ij} = 0$ for all $i > j$. Then, only 486 of the 5,196 tables remain. These 486 tables have, according to the formula for $P_0(x_{11}, \ldots, x_{46})$ given above, together a probability of 0.0441176. In order to have a total probability of 1 in the restricted sample space of the 486 possible permutations, each of these 486 permutations has the (conditional) probability $P_0(x_{11}, \ldots, x_{46})/0.0441176$.

Moreover, a further modification is required (Neuhäuser et al., 2003). For the computation of a Mann-Whitney score, we should count only observations that are possible within both groups. For example, when computing U_{23}, all broods with up to two nestlings are ignored, as these observations are not possible in a brood with three sires. With these modified Mann-Whitney scores and the restricted sample space, a permutation test can be carried out. Based on the Jonckheere-Terpstra test statistic, the resulting p-value is 0.0725.

If the usual Jonckheere-Terpstra test based on all 5,196 tables is performed, there would be a strong significance with a p-value of 0.0042. With the modified Jonckheere-Terpstra statistic T_{MJT} discussed in Section 9.2, one would get the smaller p-value of 0.0016. Of course, this modified statistic T_{MJT} can also be formed using the modified Mann-Whitney scores and, based on the restricted sample space of 486 permutations, a p-value of 0.0470 results. Tests for an umbrella alternative give even smaller p-values. These results indicate that there might be not only an inherent trend caused by the compulsory empty cells, but also a "real" trend (Neuhäuser et al., 2003).

12.2   Composite Test Statistics

In Section 3.1 a modified Lepage test was discussed that uses the BWS statistic instead of the Wilcoxon rank sum. The modified test statistic L_M does not follow any standard distribution. However, that is not relevant when applying a permutation test. The topic of this section is another test that is also composed of two statistics.

In several applications many zeros occur in addition to some positive values. For example, the amount of loss per insured person and year is often zero, because there is no loss for many insurants. If there is a loss, the amount of loss is a positive value. Lachenbruch (1976) gave another example: Cells were transplanted from one strain of mice to another. The number of cells of the new type was counted 15 days after the transplantation. In some mice, the transplant was rejected completely, and the value 0 resulted. In some other animals, the number of cells of the new type was positive. Further examples are presented by Ruxton et al. (2010).

In such a situation, some of the tests presented in previous chapters could be applied, for instance, the Wilcoxon-Mann-Whitney (WMW) test in the two-sample case (see Section 2.2). However, a more powerful test is possible when the zeros and the positive observations are analyzed separately at first.

Let us focus on the two-sample case. The number of zeros in the two groups to be compared are denoted by m_1 and m_2. In a first test the proportions of zeros are compared with the following χ^2 statistic:

$$\chi^2 = \frac{(\hat{p}_1 - \hat{p}_2)^2}{\hat{p}(1 - \hat{p})\frac{n_1 + n_2}{n_1 n_2}} \, ,$$

where the within-group proportions are estimated by $\hat{p}_i = m_i/n_i$ and $\hat{p} = (m_1 + m_2)/(n_1 + n_2)$. This test statistic is, under the null hypothesis that there is no difference between the two groups, asymptotically χ^2 distributed with one degree of freedom. If $\hat{p} = 0$ or 1, we define $\chi^2 = 0$.

In a second test one can compare the positive observations. Here, we use the WMW test based on the $n_1 - m_1$ non-zero values of group 1 and the $n_2 - m_2$ non-zero values of group 2. Let W_s denote the standardized WMW test statistic for this comparison. Under the null hypothesis, W_s^2 also has an asymptotic χ^2 distribution with 1 degree of freedom.

Lachenbruch (1976) proposed combining the two test statistics within a so-called "two-part test." To be precise, the sum $X^2 = \chi^2 + W_s^2$ is used as the new test statistic. If the two single statistics are independent, the sum X^2 would be, under the null hypothesis, asymptotically χ^2 distributed with df = 2. Lachenbruch utilizes this; he wrote that the two test statistics χ^2 and W_s are "independent under the assumption of independent errors of the binomial and continuous parts of the distribution" (Lachenbruch, 2002, p. 299). However, the generality of such independence between χ^2 and W_s seems doubtful. A permutation test with X^2, proposed by Neuhäuser et al. (2005) and Ruxton et al. (2010) and called two-part permutation test, does not need this independence. If there is no positive value within a group for any permutation, we set $W_s = 0$.

The two-part permutation test has some advantages (Neuhäuser et al., 2005; Ruxton et al., 2010): It is a valid test whether or not χ^2 and W_s are independent. No asymptotic distribution is used; thus, small samples are no problem. In addition, it reduces without any loss of power to the exact WMW test if there are positive observations only. Hence, the two-part permutation test can be used in routine analyses. Moreover, the two-part permutation test has a good power in comparison to the asymptotic test, as shown in simulation studies presented by Neuhäuser et al. (2005) and Ruxton et al. (2010).

If the two-sample test is significant, separate tests can follow in a second step of a closed testing procedure (Ruxton et al., 2010; also see the appendix). In particular for the comparison of the positive values, other test statistics can be used. Lachenbruch (1976) considered not only the WMW test, but also the parametric t test and the Smirnov test. In the two-part permutation test, one can also apply the BWS test.

12.2.1 Example

Methylation of DNA is a common method of gene regulation. The methylation patterns of tumor cells can be compared to those of normal cells; moreover, there are also differences between different types of cancer. The DNA methylation can be measured. When the tested region is not or only partially methylated, the result is negative. Such an undetectable methylation is considered a zero value. If methylation is detected, a value greater than 0 results.

Siegmund et al. (2004) compared small cell lung cancer and non-small cell lung cancer. There are forty-one observations (patients) in the first group, and forty-six in the second. Table 12.2. displays data for the gene region *MYODI*.

When comparing the negative and positive values of Table 12.2, the following 2x2 table results:

TABLE 12.2: DNA methylation data for the region *MYODI* for patients with lung cancer

Small cell lung cancer:
$(n_1 = 41)$
25 negative values
positive values: 0.21, 0.29, 0.3, 0.48,
0.5, 0.67, 1.48, 2.39, 3.49, 4.03,
6.37, 6.89, 8.21, 25.71, 33.52, 124.35
Non-small cell lung cancer:
$(n_2 = 46)$
16 negative values
positive values: 0.12, 0.13, 0.17, 0.18, 0.38,
0.46, 0.56, 0.74, 0.99, 5.14, 6.15, 7.97,
8.85, 9.83, 10.06, 14.27, 20.02, 21.43,
21.6, 27.52, 51.77, 53.32, 63.89, 67.14,
69.95, 70.78, 71.31, 79.25, 83.81, 135.7

25	16
16	30

Source Siegmund et al. 2004.

Based on these data, the χ^2 statistic is 5.969. When the WMW test is applied to the positive values, one gets $W_s^2 = 2.681$. Thus, the sum X^2 is $5.969 + 2.681 = 8.650$. The resulting asymptotic p-value based on the χ^2 distribution with df $= 2$ is 0.0132. The p-value of an approximate two-part permutation test based on 20,000 randomly selected permutation is similar: 0.0111 (Neuhäuser et al., 2005).

Chapter 13

Estimation and Confidence Intervals

As already mentioned in Chapter 1, statistical tests are the focus of this book. However, the topics of estimation and confidence intervals are also important for nonparametric models. In this chapter these topics are briefly discussed. There are several further methods not considered in this chapter, for example, the nonparametric confidence intervals for one- and two-sample problems introduced by Zhou (2005).

13.1 One-Sample Situation

A sample of independently and identically distributed random variables is denoted by $X_1, \ldots, X_n$. Let ϑ be the median of the underlying continuous distribution. Of course, the sample can consist of differences of paired observations. An estimator for ϑ is the sample median. This is the estimator that corresponds to the sign test.

For symmetric distributions the so-called Hodges-Lehmann estimator is an alternative. This estimator corresponds to the Wilcoxon signed rank test (see Hollander and Wolfe, 1999, pp. 51–55). In order to determine the Hodges-Lehmann estimator, the $M = n(n+1)/2$ means $(X_i + X_j)/2$, $i \leq j = 1, \ldots, n$, must be computed first. The median of these means is the Hodges-Lehmann estimator $\hat{\vartheta}$. In contrast to the usual procedure when applying the sign or signed rank test, zero values of X_i are considered for the estimation of ϑ.

In order to calculate a $(1 - \alpha)$ confidence interval for ϑ, the $(1 - \alpha/2)$ quantile of the permutation null distribution of the test statistic R_+ of the Wilcoxon signed rank test is required. This $(1 - \alpha/2)$ quantile is denoted by $t_{1-\alpha/2}$. Moreover, let $W_{(1)} \leq W_{(2)} \leq \cdots \leq W_{(M)}$ be the ordered means $(X_i + X_j)/2$. Then, a $(1 - \alpha)$ confidence interval for ϑ ranges from $W_{(C_\alpha)}$ to $W_{(t_{1-\alpha/2})}$, where

$$C_\alpha = \frac{n(n + 1)}{2} + 1 - t_{1-\alpha/2} \, .$$

Because the distribution of R_+ is discrete, not all confidence levels are possible. However, the coverage probability should be at least $1 - \alpha$. This

$(1 - \alpha)$ confidence interval includes exactly those values ϑ_0 for which the two-sided Wilcoxon signed rank test cannot reject the null hypothesis $\vartheta = \vartheta_0$ at level α. The equivalence between test and confidence interval is lost when some of the X_i are zero and are discarded for testing. In the case of discrete distributions, ties between the non-zero values are possible. Generalizations for those distributions can be found at Hollander and Wolfe (1999).

Let us consider Charles Darwin's data (see Table 8.1) as an example. The sample size is $n = 15$. With $\alpha = 0.05$, the $(1 - \alpha/2)$ quantile of the permutation null distribution of R_+ is $t_{1-\alpha/2} = 95$ because $P_0(R_+ \geq 95) = 0.024$ and $P_0(R_+ \geq 94) > 0.025$. Hence, $C_\alpha = 26$. Now the $M = 120$ means $(X_i + X_j)/2$ must be ordered. The median of these 120 means is $\hat{\vartheta} = (W_{(60)} + W_{(61)})/2 = (3.05 + 3.20)/2 = 3.125$, and $[W_{(26)}, W_{(95)}] = [0.45, 5.20]$ results as 95% confidence interval for ϑ.

The estimator corresponding to the sign test, that is, the median of the X_i, is 3.0. The respective confidence interval can be determined based on order statistics as follows: The coverage probability for the interval $[X_{(d)}, X_{(n+1-d)}]$ is $1 - 2P(B \leq d - 1)$, where B is binomially distributed with n and $p = 0.5$. This confidence interval can also be used for discrete distributions without exceeding the confidence level (Larocque and Randles, 2008).

This $(1 - \alpha)$ confidence interval includes exactly those values ϑ_0 for which the two-sided sign test cannot reject the null hypothesis $\vartheta = \vartheta_0$ at the significance level α. Again, the equivalence between sign test and confidence interval is lost if zeros occur and if they are discarded for testing.

For $n = 15$ and $\alpha = 0.05$, we have $1 - 2P(B \leq 3) = 0.9648$ and $1 - 2P(B \leq 4) = 0.8815$. In order to get a confidence interval with a coverage probability of at least 95%, one must choose $d = 4$. For Charles Darwin's data, the boundaries of a 95% confidence interval are the ordered values $X_{(4)} = 1.0$ and $X_{(12)} = 6.1$.

The latter confidence interval can be computed with the SAS procedure CAPABILITY and the option CIQUANTDF:

```
PROC CAPABILITY CIQUANTDF;
 VAR difference;
RUN;
```

The output displays the confidence intervals also for further quantiles:

Quantile	Estimate	95% Conf. Limits Distribution Free		-----Order Statistics-----		
				LCL Rank	UCL Rank	Coverage
100% Max	9.3					
99%	9.3	.	.	.	.	.
95%	9.3	7.0	9.3	13	15	50.05
90%	7.5	6.1	9.3	12	15	73.86
75% Q3	6.1	3.0	9.3	8	15	96.93
50% Median	3.0	1.0	6.1	4	12	96.48
25% Q1	1.0	-8.4	3.0	1	8	96.93
10%	-6.0	-8.4	1.0	1	4	73.86
5%	-8.4	-8.4	0.7	1	3	50.05
1%	-8.4	.	.	.	.	.
0% Min	-8.4					

In addition to estimators and confidence intervals, the ranks of those values that are boundaries of an interval are presented. The right column shows the coverage probabilities. This probability can be changed using the option ALPHA=; then the coverage probability is at least $1 - \alpha$. By default, $\alpha = 0.05$.

13.2 Two-Sample Situation

The two independent samples are denoted by $X_1, \ldots, X_{n_1}$ and $Y_1, \ldots, Y_{n_2}$, as in the previous chapters. Within groups, the variables X_i and Y_j, respectively, are independently and identically distributed according to the continuous distribution functions F and G. First, we consider the location-shift model $F(t) = G(t - \theta)$ for all t, and an estimator for θ is of interest. We compute all $n_1 n_2$ differences $Y_j - X_i$, $j = 1, \ldots, n_2$, $i = 1, \ldots, n_1$. Then, an estimator for θ is the median of these $n_1 n_2$ differences. This estimator $\hat{\theta}$ is also called the Hodges-Lehmann estimator (Hollander and Wolfe, 1999, p. 125).

Let us consider the example of a clinical trial described in Section 7.2. With $n_1 = n_2 = 10$ there are 100 differences $Y_j - X_i$. When the observations in the placebo group are denoted by Y_j, $\hat{\theta} = -17.5$ is the Hodges-Lehmann estimator.

In order to get a symmetric $(1 - \alpha)$ confidence interval for θ, we require the $(1 - \alpha/2)$ quantile of the distribution of the Wilcoxon rank sum under H_0: $\theta = 0$, this quantile is denoted here by $w_{1-\alpha/2}$. Moreover, we define

$$C_\alpha = \frac{n_1(2n_2 + n_1 + 1)}{2} + 1 - w_{1-\alpha/2}.$$

The confidence interval is also based on the $n_1 n_2$ differences $Y_j - X_i$. The lower bound of the $(1 - \alpha)$ confidence interval is the value of that difference that has the position C_α in the ordered list of all $n_1 n_2$ differences. The upper bound is the difference at position $n_1 n_2 + 1 - C_\alpha$. This confidence interval according to Moses corresponds to the Wilcoxon rank sum test. The $(1 - \alpha)$ confidence interval includes exactly those values θ_0 for which the two-sided hypothesis $\theta = \theta_0$ cannot be rejected at level α (Hollander and Wolfe, 1999, pp. 132–133).

For the example of the clinical trial (Table 7.2), we have $n_1 = n_2 = 10$, $\alpha = 0.05$, and therefore $w_{1-\alpha/2} = w_{0.975} = 132$. Hence, $C_{0.05} = 24$ and $n_1 n_2 + 1 - C_{0.05} = 77$, and $[-31, -1]$ is the resulting 95% confidence interval for θ. Due to the discreteness of the rank sum, the coverage probability is not exactly equal to 95%; instead, it is $\geq$ 95%. This confidence interval is problematic because the location-shift model seems not to be appropriate for these data (see Section 7.2).

In particular when the variability differs between the two groups to be compared, one may consider the relative effect. It was mentioned in Section 3.2 that

$$\hat{p} = \frac{1}{N} \left(\bar{R}_2 - \bar{R}_1 \right) + 0.5$$

is an unbiased estimator for the relative effect p (Brunner and Munzel, 2000). The corresponding $(1 - \alpha)$ confidence interval is

$$\hat{p} \pm \frac{t_{\mathrm{df}, 1-\alpha/2}}{n_1 n_2} \sqrt{n_1 \tilde{S}_1^2 + n_2 \tilde{S}_2^2},$$

where the t quantile with df degrees of freedom is denoted by $t_{df, 1-\alpha/2}$, the variance estimators can be computed according to the formulas given in section 3.2. For large sample sizes one can also use the $(1 - \alpha/2)$ quantile of the standard normal distribution (Brunner and Munzel, 2002, Section 2.1).

Our example is still the clinical trial presented in Section 7.2. Based on these data, the estimated relative effect is $\hat{p} = 0.78$. The variance estimators $\tilde{S}_i^2$ are 11.29 and 1.73, and there are df = 11.70 degrees of freedom. Hence, the 95% confidence interval based on the t distribution is

$$0.78 \pm \frac{2.1850}{100} \sqrt{10 \cdot 11.29 + 10 \cdot 1.73} = [0.531, 1.029].$$

The relative effect p is a probability and therefore must be a value between 0 and 1. However, in this example the upper bound of the confidence interval is 1.029. For such a situation, Brunner and Munzel (2002, pp. 83–84) recommended applying the so-called δ method with the transformation $g(\hat{p}) = \ln(\hat{p}/(1 - \hat{p}))$. For $g(p)$ the boundaries based on the t quantiles are

$$p_U^g = \ln\left(\frac{\hat{p}}{1 - \hat{p}} \right) - \frac{\hat{\sigma}_N \cdot t_{\mathrm{df}, 1-\alpha/2}}{\hat{p}(1 - \hat{p})\sqrt{N}} \quad \text{and}$$

$$p_O^g = \ln\left(\frac{\hat{p}}{1-\hat{p}}\right) + \frac{\hat{\sigma}_N \cdot t_{\mathrm{df},1-\alpha/2}}{\hat{p}(1-\hat{p})\sqrt{N}} \,,$$

where

$$\hat{\sigma}_N^2 = \frac{N}{n_1 n_2} \sum_{i=1}^{2} \frac{\tilde{S}_i^2}{N-n_i} \,.$$

Through back transformation, the resulting $(1-\alpha)$ confidence interval for the relative effect p is

$$\left[\frac{\exp(p_U^g)}{1+\exp(p_U^g)} \,, \quad \frac{\exp(p_O^g)}{1+\exp(p_O^g)}\right] \,.$$

In the example we have $p_U^g = -0.187$ and $p_O^g = 2.719$; therefore, [0.453, 0.938] is the 95% confidence interval for the relative effect. Confidence intervals for relative effects can also be presented when more than two groups are compared (see Brunner and Munzel, 2002, Section 2.2.6).

For small sample sizes, the confidence interval should be based on a permutation distribution. This is possible using the general principle of Bauer (1972). The values of one group must be shifted by θ, that is, for example $Y_j - \theta$ is computed for all values of group 2. Then, the $(1-\alpha)$ confidence interval includes the values of θ for which the corresponding test cannot reject the null hypothesis at level α. Details as well as calculations for an example are presented by Neubert (2006).

13.3 Bootstrap and Jackknife

The principle of the bootstrap can be used for testing, but also for point and interval estimation. Let us assume that a sample $x_1, \ldots, x_n$ of size n is observed; the underlying distribution is unknown. A parameter θ is estimated by $\hat{\theta} = \hat{\theta}(x_1, \ldots, x_n)$. The bootstrap was introduced in 1979 in order to estimate the standard error of the estimator $\hat{\theta}$ (Efron and Tibshirani, 1993, p. 45). If $\hat{\theta} = \bar{X}$, the square root of $\sum_{i=1}^{n}(x_i - \bar{x})^2/n(n-1)$ is an estimator of the standard error. However, this estimator cannot be applied when θ is estimated, for example, using a median. Through bootstrap or jackknife it is possible to estimate the standard error, or the variance, of $\hat{\theta}$ (Efron, 1982, pp. 1–2). For the bootstrap, the following algorithm can be applied:

(a) Compute the estimator $\hat{\theta}$ for the observed values.

(b) Draw B independent bootstrap samples of size n with replacement from the n values $x_1, \ldots, x_n$.

(c) Calculate the estimator $\hat{\theta}$ for every bootstrap sample; these estimators are denoted by $\hat{\theta}_1^*, \ldots, \hat{\theta}_B^*$.

(d) Estimate the variance of $\hat{\theta}$ by the empirical variance of the B bootstrap estimators $\hat{\theta}_1^*, \ldots, \hat{\theta}_B^*$; thus

$$\widehat{Var}(\hat{\theta}) = \frac{1}{B-1} \sum_{b=1}^{B} \left(\hat{\theta}_b^* - \hat{\theta}^* \right)^2$$

with $\hat{\theta}^* = \sum_{b=1}^{B} \hat{\theta}_b^*/B$.

If the sample size n is very small it may be possible to consider all $\binom{2n-1}{n}$ possible bootstrap samples instead of a random sample of B bootstrap samples.

Of course, the principle used in the above algorithm can also be used to estimate parameters other than the variance. For instance, Efron and Tibshirani (1993, pp. 49–50) discussed the estimation of the coefficient of correlation. Because the estimation is based on the n observations and therefore on the empirical distribution function, the method is called a nonparametric bootstrap estimation. For a parametric bootstrap, additional assumptions are required; see Efron and Tibshirani (1993) for details.

The bootstrap can also be used for confidence intervals. There are different methods (see Efron and Tibshirani, 1993); here, the so-called percentile method is presented for illustration. Let $\hat{\theta}_B^{*(\alpha)}$ be the 100αth empirical percentile (α quantile) of the bootstrap estimators $\hat{\theta}_1^*, \ldots, \hat{\theta}_B^*$. In the case of, for example, $B = 2000$ and $\alpha = 0.05$, $\hat{\theta}_B^{*(\alpha)}$ is the $B\alpha = 100$th ordered value of the B bootstrap estimators $\hat{\theta}_1^*, \ldots, \hat{\theta}_B^*$. Accordingly, $\hat{\theta}_B^{*(1-\alpha)}$ is the $100(1-\alpha)$th empirical percentile. The $(1-2\alpha)$ bootstrap confidence interval according to the percentile method is then $\left[\hat{\theta}_B^{*(\alpha)}, \hat{\theta}_B^{*(1-\alpha)} \right]$. If $B\alpha$ is not an integer, Efron and Tibshirani (1993, pp. 160–170) suggested defining the empirical α and $(1-\alpha)$ quantiles by the k-th largest and $(B+1-k)$-th largest values of the B bootstrap estimators, where k is the largest integer $\leq (B+1)\alpha$. Moreover, we assume $\alpha \leq 0.5$.

Like the bootstrap, the so-called jackknife is a resampling method. When applying the jackknife, there is no drawing of samples; instead, single observations are eliminated. The jackknife was proposed in the mid-20th century. For details the reader is referred to Efron (1982) and Efron and Tibshirani (1993, Chapter 11).

Let us assume that the random variables $X_1, X_2, \ldots, X_n$ are independently and identically distributed according to the distribution function F. A parameter θ is estimated by $\hat{\theta} = \hat{\theta}(X_1, X_2, \ldots, X_n)$. Obviously,

$$\hat{\theta}_{-i} = \hat{\theta}(X_1, \ldots, X_{i-1}, X_{i+1}, \ldots, X_n)$$

is also an estimator for θ. The statistics

$$J_i(\hat\theta) = \hat\theta + (n-1)\left(\hat\theta - \hat\theta_{-i}\right), \quad i = 1, \ldots, n,$$

are called pseudo-values and

$$J(\hat\theta) = \frac{1}{n}\sum_{i=1}^{n} J_i(\hat\theta)$$

is the jackknife estimator for θ based on $\hat\theta$.

This jackknife estimator $J(\hat\theta)$ is often less biased than the estimator $\hat\theta$, which was used as the basis for the jackknife. As an example we consider the estimation of the variance. As the basis, we apply the (biased) variance estimator

$$\hat\theta = \frac{1}{n}\sum_{i=1}^{n}(X_i - \bar{X})^2 \ .$$

The resulting jackknife estimator is the usually applied, unbiased variance estimator

$$J(\hat\theta) = \frac{1}{n-1}\sum_{i=1}^{n}(X_i - \bar{X})^2 \ ,$$

(Efron and Tibshirani, 1993, p. 151). In general,

$$\frac{n-1}{n}\sum_{i=1}^{n}\left(\hat\theta_{-i} - \hat\theta_{(.)}\right)$$

is an estimator for the variance of $\hat\theta$, where $\hat\theta_{(.)}$ is the mean of $\hat\theta_{-1}, \ldots, \hat\theta_{-n}$ (Efron, 1982, p. 13).

In a generalized jackknife, blocks of observations are eliminated rather than single values (Efron, 1982, p. 7). The jackknife principle can also be used to carry out statistical tests (Rodgers, 1999). However, such an application of the jackknife is not common in statistical practice, and therefore not discussed here.

Appendix

The first three parts of this appendix present some basic ideas on the topics level of measurement, statistical tests, and multiple testing. These are short summaries; for further details it is referred to introductory books such as Siegel (1956) and Mood et al. (1974) for the topics level of measurement and statistical tests, respectively, or to review articles such as Pigeot (2000) or Neuhäuser (2006b) for multiple tests. The last part of this appendix provides R code for several nonparametric tests presented in the previous chapters.

A.1 Level of Measurement

The weakest level of measurement is the nominal scale. A variable is nominal when the categories to which different objects belong can be identified, but the categories cannot be ordered. Examples include gender, nationality, or blood group.

Variables at the ordinal scale are qualitative, as are nominal ones. The difference is that ordinal variables can be ordered. Thus, one category is greater than another, but the difference between categories cannot be interpreted. Hence, no information is lost when applying any transformation that does not change the order of the categories. Examples are quality categories and school grades. A further example is the system of grades in the military services.

Values of a quantitative, or metric, variable are numbers obtained by, for example, measuring, weighing, or counting. Now, there is not only an order relation, but the distances between any two numbers are of known size. Hence, differences between two values can be computed. There are discrete and continuous quantitative variables. A continuous variable such as body weight can take on any value (within an interval). In contrast, a discrete variable such as the number of children can take only values of a countable set.

Quantitative variables can be further classified; there are the interval scale and the ratio scale. At the interval scale, differences are interpretable, and the distances between all objects are known. However, a ratio between two values cannot be interpreted because the zero point is arbitrary. An example is the temperature measured in centigrade or Fahrenheit. At the ratio scale there is a true zero point and ratios are also meaningful. The ratio of any two values is independent of the used unit of meaurement. Examples are body weight and body height.

A variable with a high level of measurement can be transformed to a weaker level. For instance, the continuous variable body height can be binned into the three categories: small, medium, and large. Then the new transformed variable has an ordinal scale.

A.2   Statistical Tests

A statistical test can decide between two statements that are called hypotheses. There are the so-called null hypothesis H_0 and the alternative (alternative hypothesis) H_1. The intersection of these two hypotheses is empty. With the statistical test one can decide to maintain or to reject the null hypothesis. In total, four combinations are possible:

	Decision for	
True is	Null hypothesis	Alternative
Null hypothesis	correct	Type I error
Alternative	Type II error	correct

Two out of these four decisions are wrong. If the null hypothesis is rejected although it is true, the error is called a type I error. A type II error occurs if the null hypothesis is not rejected although it is not true. In general, it is not possible to minimize the probabilities for both types of error simultaneously. If the probability for one error is reduced, the probability for the other error increases. Therefore, the probability for the type I error, that is, the type I error rate, is usually limited to α. Under that constraint the test that minimizes the type II error rate is chosen for a specific test problem.

The limit α is called the significance level of the test. When a test statistic is selected, a rejection region can be specified. If the rejection region includes the value of the test statistic calculated with the actually observed data, the null hypothesis can be rejected. Therefore, the rejection region contains values that provide evidence against the null hypothesis, that is, values far away from the—under the null hypothesis—expected value of the test statistic.

In order to control the significance level α, the conditional probability that the value of the test statistic is within the rejection region, under the assumption that the null hypothesis is true, must not be larger than α. In the case of a continuously distributed test statistic, it is possible that this conditional probability is exactly equal to α. That is in general not possible in case when the test statistic has a discrete distribution. The maximum conditional probability to reject the null hypothesis, under the assumption that the null hypothesis is true, is called actual level, or size, of the test. The prespecified significance level α is the nominal level. If the actual level is smaller than α, the test is conservative; if the actual level is larger than α, the test is anticonservative, or liberal.

The nominal level must be specified before the test is performed. The usual value is $\alpha = 0.05$. This choice is arbitrary, but very common. In the case of a one-sided test, an α of 0.025 is sometimes used.

In contrast to α there usually is no control about the probability for a type II error. Therefore, no conclusion is possible when the null hypothesis cannot be rejected. In case of doubt, a statistical test decides in favor of the null hypothesis H_0; thus, one cannot prove H_0. In contrast, one can statistically prove the alternative H_1: If the null hypothesis is rejected, the possible error is limited by α. The result is called significant if the null hypothesis is rejected.

The probability of rejecting a wrong null hypothesis is called the power of the test. The power is not constant; it depends on the considered point of the alternative. Furthermore, the power depends on α and sample size, among other things.

Statistical software systems usually present a p-value. Let T be a test statistic that, without loss of generality, tends to larger values under the alternative. The null hypothesis can be rejected if

$$t(x) \geq c_\alpha$$

holds for the actually observed value $t(x)$ of the test statistic, and c_α denotes the lower boundary of the rejection region. For a simple null hypothesis, the p-value is defined as $P_0(T \geq t(x)|H_0)$, that is, the probability for a test statistic that is at least as large as actually observed based on the data x, under the assumption that the null hypothesis H_0 is true. Note that we assume that the larger the value of the test statistic, the more evidence exists against the null hypothesis.

In general, the null hypothesis H_0 can be rejected if the p-value is less than or equal to α. If the p-value is greater than α, H_0 cannot be rejected. Hence, the p-value is the smallest level at which the test is just significant.

A.3   Multiple Testing

Multiplicity of statistical inference can arise in a variety of settings. Examples are studies where more than one variable is analyzed, and studies with more than two groups. In the latter case, a first test, for instance a Kruskal-Wallis test, might demonstrate a difference between groups. The multiple testing occurs when after a first significant test, pairwise comparisons are performed in order to identify differences. There is extensive literature on multiple testing. Here, we briefly describe some basic concepts (based on the review of Neuhäuser, 2006b)

It is well known that the probability of obtaining significant results, just by chance, increases considerably with the overall number of statistical tests carried out. Therefore, it might be not appropriate to declare a p-value significant just because it is not larger than α.

An appropriate multiple testing procedure should control the rate of false positive conclusions at an acceptable level. A multiple testing procedure strongly controls the type I error rate (i.e., the probability of false positive decisions) when the probability of erroneously rejecting any configuration of subhypotheses is controlled by α, although some subhypotheses may be true. To be precise, strong control means that the probability of rejecting at least one true null hypothesis is not larger than α, irrespective of which and how many individual null hypotheses are true.

Let us assume a collection of k tests that will be carried out simultaneously. The simplest multiplicity adjustment, that is, the classical Bonferroni technique, compares the p-values of each single test with the adjusted level α/k rather than α. For instance, in case of five tests and the usual 5% level, the new threshold is $0.05/5 = 0.01$.

The Bonferroni-Holm sequentially rejective procedure provides a more powerful test. First, the p-values of the k tests must be ordered from the smallest $p_{(1)}$ to the largest p-value $p_{(k)}$, that is, $p_{(1)} \leq p_{(2)} \leq \cdots \leq p_{(k)}$. If the smallest p-value $p_{(1)}$ takes on a value $\leq \alpha/k$, then the corresponding null hypothesis can be rejected. With regard to this single test, there is no difference between the Bonferroni and the Bonferroni-Holm methods. If $p_{(1)} > \alpha/k$, the procedure stops and all tests are nonsignificant. In contrast, if $p_{(1)} \leq \alpha/k$, the second-smallest p-value $p_{(2)}$ is compared with $\alpha/(k-1)$. If $p_{(2)} > \alpha/(k-1)$, the corresponding test and all other tests with larger p-values do not reject their null hypotheses. However, if $p_{(2)} \leq \alpha/(k-1)$, the corresponding test is significant, and the procedure proceeds until $p_{(i)} > \alpha/(k-i+1)$, or until the last null hypothesis is rejected in the case of $p_{(k)} \leq \alpha$.

Note that all null hypotheses that are rejected using the Bonferroni adjustment are also rejected with the Bonferroni-Holm procedure. The latter method, however, may reject additional null hypotheses. Both methods, Bonferroni and Bonferroni-Holm, strongly control the type I error rate.

If the overall decision rule is that a significance is needed regarding all k individual null hypotheses, no multiplicity adjustment such as the Bonferroni-Holm method is needed. All tests can be performed with the unadjusted level α, but all single p-values must be less than or equal to α. This procedure inflates the type II error rate, which must be taken into account for sample size determination. Similarly, an adjustment of the significance level is not necessary when multiple hypotheses are to be tested in a prespecified order according to their relevance. Then each hypothesis is tested at full level α in the preassigned order, as long as the prior null hypotheses have been rejected.

When a closed testing procedure is applied, all tests are performed at level α, but additional tests might be necessary. Within such a procedure the set of null hypotheses to be tested must be closed under intersection; that is, the intersection of any two of the null hypotheses must be an element of the set of null hypotheses. In order to reach closeness, additional hypotheses often must be included. A null hypothesis H_0^i can be rejected if it is rejected in the corresponding test at level α, and if all other null hypotheses containing H_0^i are also rejected in their level α tests.

A closed testing procedure can be performed in a stepdown manner. It starts with testing the intersection of all null hypotheses, that is, the global null hypothesis. If this global null hypothesis is rejected at level α, the tests proceed to the intersection null hypotheses one stage lower.

Let us, consider a situation with the three null hypotheses H_0^1, H_0^2, and H_0^3. We further assume that $H_0^3 = H_0^1 \cap H_0^2$. Then, the set of these three null hypotheses is closed under intersection, and no further null hypothesis must be included. The global null hypothesis is H_0^3. When H_0^3 can be rejected at level α, the two other null hypotheses H_0^1 and H_0^2 can be tested, both with the unadjusted level α. Further, more complex examples can be found in Kropf (1988). The closed testing procedure strongly controls the type I error rate. For example, the Bonferroni-Holm method is a special case of a closed test.

When performing a multiplicity adjustment, or when applying a closed testing procedure, the power for an individual test may become very low, in particular when the number of tests is large. Consequently, relatively strong effects are only likely to be recognized as significant. That might be too stringent, for example in purely exploratory studies.

An alternative is the control of the so-called false discovery rate. When many tests are performed, keeping the proportion of type I errors ("false discoveries") at a low level might be an alternative to controlling the chance of making even a single type I error. The control of the false discovery rate (FDR) means that the expected proportion of "number of type I errors/number of significant tests" is maintained at a desired level. This approach was developed by Benjamini and Hochberg (1995), and is, for instance, widely adopted in genomic research.

A.4 Nonparametric Tests in R

In the different chapters of this book it has been shown how the different tests can be carried out using SAS. Of course, there are several other statistical software systems. Many permutation tests can be performed using StatXact. Here, we show how R can be used for nonparametric tests. More general details about statistical analyses using R are provided, for example, by Everitt and Hothorn (2006). Duller (2008) presents R code for several nonparametric methods.

The Fisher-Pitman permutation (FPP) test for two samples and the Wilcoxon-Mann-Whitney (WMW) test can be performed using functions of the package `exactRankTests` or the more recent package `coin` (abbreviation for conditional inference). The following code analyzes the example data set of Good (2001) discussed in Sections 2.1 and 2.2 with the FPP and the WMW tests:

```
library(exactRankTests)
good <- data.frame(treated=c(90,110,118,121),
   untreated=c(12,22,34,95))
perm.test(good$treated, good$untreated, exact=TRUE)
wilcox.exact(good$treated, good$untreated, exact=TRUE,
   paired=FALSE)
```

The function `perm.test()` invokes the FPP test, the used test statistic is the sum of observations in one sample. The function `wilcox.exact()` invokes the WMW test, the test statistic displayed in the output (not shown) is the Mann-Whitney statistic U. Both functions are part of the package `exactRankTests`. Note that the function `wilcox.test()` from the standard package `stats` can perform an exact WMW test too, but only if there are no ties. The function `wilcox.exact()` can handle ties; however, the package `exactRankTests` is required for this function. This package uses the shift algorithm of Streitberg and Röhmel (1987); details are presented by Hothorn and Hornig (2002).

Both tests, FPP and WMW, can also be carried out using the package `coin`; in this case, the standardized statistics are displayed in the output (not shown):

```
library(coin)
good <- data.frame(y=c(90,110,118,121,12,22,34,95),
   treat=factor(rep(1:0,rep(4,2))))
oneway_test(y ~ treat, data=good, distribution="exact")
wilcox_test(y ~ treat, data=good, distribution="exact")
```

The package `coin` is a very flexible tool; more details are presented in Everitt and Hothorn (2006, Chapter 3) as well as Hothorn et al. (2006). The Van Elteren test (see Section 11.1) is a stratified WMW test and, hence, can also be performed using the `coin` infrastructure. This package also includes the function `ansari_test()` that performs the Ansari-Bradley test, a part of Lepage's location-scale test (see Section 3.1). Moreover, there is also the function `ansari.test()` for the Ansari-Bradley test. The latter function does not require the package `coin`. Lepage's location-scale test can be directly applied using the R code available at `www.rheinahrcampus.de/fileadmin/fb_mathematik_technik/neuhaeuser/ Lepage_Test_R.pdf`.

The following code defines a function called `BWS.Test()` that can carry out the Baumgartner-Weiss-Schindler (BWS) test (see Section 2.3). When choosing the option `exact=TRUE`, an exact permutation test is performed, while `exact=FALSE` invokes an approximate permutation test with M randomly selected permutations. Moreover, an approximate test is applied when the total sample size is too large. To be precise, the approximate test is used when the total sample size is larger than twenty. However, that value can be changed within the code given below. Of course, this code can be modified in order to perform permutation tests with other test statistics.

```
BWS.Test <- function(X1, X2, M=10000, exact=FALSE){
Pvall <- function(X1, X2, n, m, N){
   Z=c(X1,X2);
   RZ=rank(Z, na.last = TRUE, ties.method ="average");
   R1=RZ[1:n];
   R2=RZ[(n+1):N];
   R1=sort(R1);
   R2=sort(R2);

   i=1:n;
   j=1:m;
   Bx=(1/n)*sum((R1-(N/n)*i)^2/( (i/(n+1))*(1-(i/(n+1)))*
      ((m*N)/n) ));
   By=(1/m)*sum((R2-(N/m)*j)^2/( (j/(m+1))*(1-(j/(m+1)))*
      ((n*N)/m) ));
   B=0.5*(Bx+By);   # test statistic

   Out=B;
   } # end of function pvall

Z=c(X1,X2);
twert=0;
```

```
n=length(X1);        # sample sizes
m=length(X2);
N=n+m;

if(exact==TRUE & N>20){
   gg=c("There are too many permutations for exact test: ",
      choose(N,n));
   print(gg,quote = FALSE);
   print("The test is performed as an approximate permutation
      test." , quote = FALSE);
   exact=FALSE;
}

testimate=Pvall(X1, X2, n, m, N);
anzPerm=gamma(N+1)/(gamma(n+1)*gamma(N-n+1))
pval=0

# all permutations
if(exact==TRUE){
   k=n;
   P=array(0, c(N, 1));

   for(i in 1:N)
   {
     P1=P;
     P2=P;
     P1[N-i+1,1:dim(P)[2]]=1;
     P2[N-i+1,1:dim(P)[2]]=0;
     P=array(c(P1, P2), c(N, (2*dim(P)[2])));
   } #end for

   Sums=apply(P, 2, sum);
   kpers=1:length(Sums);
   kpers=kpers[Sums==k]
   P=P[1:N,  kpers];

   Z1=array(Z, c(N, dim(P)[2]));
   Z1=Z1[P==1];
   perm1=array(Z1, c(k, dim(P)[2]));
   Z2=array(Z, c(N, dim(P)[2]));
   Z2=Z2[P==0];
   perm2=array(Z2, c(N-k, dim(P)[2]));
```

```
  for(i in 1:dim(perm1)[2])  # calling the function Pvall
  {
    out=Pvall(perm1[1:n, i], perm2[1:m, i], n, m, N);
    if(out>=testimate){
      pval=pval+1
    }
  }  #end for
  pval=pval/anzPerm
} # end if

# random sample of permutations
if(exact==FALSE){
  for(i in 1:M){
    Pe=rep(0, N);
    while(sum(Pe) < n){
      r=floor(runif(1, min=1, max=(N+1)));
      Pe[r]=1;
    } #end while
    perm1=Z[Pe==1];
    perm2=Z[Pe==0];

    out=Pvall(perm1, perm2, n, m, N);
                        # calling the function Pvall
    if(out>=testimate){
      pval=pval+1
    }
  }
  pval=pval/M
} # end if

print("Two sample BWS permutation test",quote = FALSE);
print("H0 = No difference between groups",quote = FALSE);

if(exact==FALSE){
  gg=c("Number of randomly selected permutations: M=",M);
  print(gg,quote = FALSE);
}
if(exact==TRUE){
  gg=c("Number of permutations: ",choose(N,n));
  print(gg,quote = FALSE);
}

pval=round(pval, digits = 4)
```

```
gg=c("p-value =", pval);
print(gg,quote = FALSE);
testimate=round(testimate, digits = 4)
gg=c("B =", testimate);
print(gg,quote = FALSE);
print("",quote = FALSE);


Allout=pval;
}
```

After defining the function `BWS.Test()`, the BWS test can be computed as follows; again the example data of Good (2001) are used:

```
X1 <- c( 90,110,118,121)
X2 <- c(12,22,34,95)
BWS.Test(X1, X2, M=10000, exact=TRUE)
```

The Brunner-Munzel test (see Section 3.2) can be carried out as a permutation test with R code available at www.biostat.uni-hannover.de/staff/neuhaus/BMpermutation_test.txt (Neuhäuser and Ruxton, 2009b).

R code for bootstrap tests (see Section 3.3) can be found in Good (2005); that book also provides R code for approximate permutation tests. R code for permutation tests, especially for complex designs, is also given in Basso et al. (2009). Wilcox (2003) presents R code for the tests discussed in his book, including code for the Brunner-Munzel test as well as for bootstrap and permutation tests.

The FPP test after the Levene transformation, proposed in Section 3.4, can also be performed using the function `oneway_test` of the package `coin`. Here, the example data set of colony sizes in bumblebees (see Table 3.2) is used:

```
library(coin)
bee <- data.frame(size=c(14,24,26,98,12,105,85,
    40,14,18,28,11,39,17,37,52,30,65,35),
    colony=factor(rep(1:0,c(7,12))))
bee.trans <- data.frame(bee, levene = abs(c(
    bee$size[bee$colony==1] - median(bee$size[bee$colony==1]),
    bee$size[bee$colony==0] - median(bee$size[bee$colony==0]))))
oneway_test(levene ~ colony, data=bee.trans,
    distribution="exact")
```

For the (Kolmogorov-)Smirnov test (see Chapter 4), there is the function `ks.test()`. With this function, however, it is not possible to carry out an exact test in case of ties. Exact p-values are not available, either, for the one-sided two-sample case. An alternative way is to modify the above-mentioned code for the BWS test in order to perform the exact permutation Smirnov test.

The sign test (see Section 8.1) can be applied using the function `binom.test()` designed for the exact binomial test. Charles Darwin's data (see Table 8.1) can be analyzed as follows:

```
darwin.binary <- data.frame(neg=2, all=15)
binom.test(darwin.binary$neg, darwin.binary$all)
```

For the exact Wilcoxon signed rank test (see Section 8.2), the function `wilcox.test()` can be used if there are no ties. However, ties frequently occur; therefore, we suggest using the function `wilcoxsign_test()` of the package coin. Again, Darwin's data are analyzed:

```
library(coin)
darwin <- data.frame(w1=c(23.5,12.0,21.0,22.0,19.1,21.5,
    22.1,20.4,18.3,21.6,23.3,21.0,22.1,23.0,12.0),
    w2=c(17.4,20.4,20.0,20.0,18.4,18.6,18.6,15.3,
    16.5,18.0,16.3,18.0,12.8,15.5,18.0))
wilcoxsign_test(darwin$w1 ~ darwin$w2, distribution=exact(),
    zero.method="Wilcoxon")
```

Again, the standardized test statistic is displayed in the output (not shown). The asymptotic test is carried out if `distribution=exact()` is omitted. The function `wilcoxsign_test()` can also apply Pratt's modification (see Section 8.2). In order to use this modification, `zero.method="Wilcoxon"` may be replaced by `zero.method="Pratt"`. Please note that Pratt's method is applied by default.

The asymptotic McNemar test (see Section 8.5) can be performed using the function `mcnemar.test()`, here for the data set of Table 8.3. In order to get the result as presented in Section 8.5, one must specify `correct=FALSE` to avoid a continuity correction:

```
smoke<-matrix(c(132,21,49,98),nrow=2)
mcnemar.test(smoke, correct=FALSE)
```

When the McNemar test should be applied as an exact test, one needs the function `mcnemar.exact()` of the package `exact2x2`:

```
library(exact2x2)
mcnemar.exact(smoke)
```

The function `kruskal.test()` can be used to carry out the asymptotic Kruskal-Wallis test (see Section 9.1). The following code invokes this test for the data set of Table 9.1:

```
motivation <- data.frame(value=c(9, 11, 7, 8, 9, 9, 10,
    11, 12, 8, 9, 10, 11, 10, 12, 13, 11, 9, 10, 12, 12,
    15, 17, 15, 10, 16, 14, 12),
    group=factor(rep(1:4,rep(7,4))))
kruskal.test(value ~ group, data=motivation)
```

The Kruskal-Wallis test is possible as an approximate permutation test based on B permutations with the function `kruskal_test()` of the package coin:

```
library (coin)
kruskal_test(value ~ group, data=motivation,
    distribution=approximate(B=10000))
```

After a significant Kruskal-Wallis test, pairwise WMW tests can be performed. This is directly possible using the function `pairwise.wilcox.test()`. This function enables pairwise comparisons between group levels with a correction for multiplicity. For example, the Bonferroni or Bonferroni-Holm adjustment can be chosen.

The Jonckheere-Terpstra trend test (see Section 9.2) can be performed as an asymptotic test using the function `JT.test()` of the bioconductor's package SAGx. For an exact permutation test, the function `jonckheere.test()` of the package `clinfun` can be applied. Here, we use the example data of Table 9.2:

```
library(clinfun)
mutation <- data.frame(number=c(101, 117, 111,  91,  90, 107,
    103, 133, 121, 136, 140, 144, 190, 161, 201, 146, 120, 116),
    dosis=ordered(rep(c(0,10,30,100,300,1000),rep(3,6))))
jonckheere.test(mutation$number, mutation$dosis,
    alternative = "increasing"))
```

To apply the Friedman test (see Section 9.3), there is the function `friedman.test()`. Moreover, within the package coin, the function `friedman_test()` is available.

The χ^2 test (see Section 10.1) can be carried out using the function `chisq.test()`. This is illustrated here with the data of Table 10.2:

```
x <- matrix(c(26, 15, 23, 10), nrow = 2)
chisq.test(x, correct=FALSE)
```

By default, R applies a continuity correction. Here, no correction is performed because of the option `correct=FALSE`. An approximate χ^2 test can also be performed using the function `chisq.test()`:

```
chisq.test(x, correct=FALSE, simulate.p.value=TRUE, B=10000)
```

No continuity correction is applied in the case of `simulate.p.value=TRUE`; B again specifies the number of replicates for the approximate test.

Pearson's correlation coefficient (see Section 10.3) can be computed using the function `cor()`. An asymptotic test is additionally performed when using the function `cor.test()`, as illustrated here for the data set of Table 10.3:

```
x <- c(1,60,124,223,303)
y <- c(31.25,28.71,23.67,18.13,15.53)
cor.test(x,y)
```

If Spearman's rank correlation is appropriate, the option `method="spearman"` must be added.

Bibliography

Agresti A (2003): Dealing with discreteness: making 'exact' confidence intervals for proportions, differences of proportions, and odds ratios more exact. *Statistical Methods in Medical Research* 12, 3–21.

Anderson MJ (2001): Permutation tests for univariate or multivariate analysis of variance and regression. *Canadian Journal of Fisheries and Aquatic Sciences* 58, 626–639.

Ansari AR and Bradley RA (1960): Rank-sum tests for dispersion. *Annals of Mathematical Statistics* 31, 1174–1189.

Armitage P (1955): Tests for linear trends in proportions and frequencies. 11, 375–386.

Baer B and Schmid-Hempel P (1999): Experimental variation in polyandry affects parasite loads and fitness in a bumble-bee. *Nature* 397, 151–154.

Basso D, Pesarin F, Salmaso L and Solari A (2009): *Permutation tests for stochastic ordering and ANOVA*. Springer, Dordrecht.

Bauer DF (1972): Constructing confidence sets using rank statistics. *Journal of the American Statistical Association* 67, 687–690.

Bauer P, Brannath W and Posch M (2001): Flexible two-stage designs: an overview. *Methods of Information in Medicine* 40, 117–121.

Bauer P and Köhne K (1994): Evaluation of experiments with adaptive interim analyses. *Biometrics* 50, 1029–1041.

Baumgartner W, Weiss P and Schindler H (1998): A nonparametric test for the general two-sample problem. *Biometrics* 54, 1129–1135.

Behnen K and Neuhaus G (1989): *Rank tests with estimated scores and their applications*. Teubner, Stuttgart.

Bender R, Lange S and Ziegler A (2007): Wichtige Signifikanztests. *Deutsche Medizinische Wochenschrift* 132, e24–e25.

Benjamini Y and Hochberg Y (1995): Controlling the false discovery rate—a practical and powerful approach to multiple testing. *Journal of the Royal Statistical Society* B 57, 289–300.

Berger VW (2000): Pros and cons of permutation tests in clinical trials. *Statistics in Medicine* 19, 1319–1328.

Berger VW (2001): The p-value interval as an inferential tool. *The Statistician* 50, 79–85.

Berger VW (2009): A Socratic dialogue. *Journal of Modern Applied Statistical Methods* 8, 316–321.

Berger VW and Ivanova A (2002): The bias of linear rank tests when testing for stochastic order in ordered categorical data. *Journal of Statistical Planning and Inference* 107, 237–247.

Berger VW, Matthews JR and Grosch EN (2008): On improving research methodology in clinical trials. *Statistical Methods in Medical Research* 17, 231–242.

Berger VW and Zhou Y (2005): Kolmogorov-Smirnov tests. In: Everitt BS and Howell DC (Eds.): *The encyclopedia of statistics in behavioral science*, vol. 2. Wiley, Hoboken, pp. 1023–1026.

Bergmann R, Ludbrook and Spooren WPJM (2000): Different outcomes of the Wilcoxon-Mann-Whitney test from different statistics packages. *American Statistician* 54, 72–77.

Berry JJ (1995a): A simulation-based approach to some nonparametric statistics problems. *Observations* 5, 19–26.

Berry JJ (1995b): Obtaining exact significance levels for various nonparametric two-independent-samples problems. *Observations* 5, 40–52.

Berry KJ, Mielke PW and Mielke HW (2002): The Fisher-Pitman permutation test: An attractive alternative to the *F* test. *Psychological Reports* 90, 495–502.

Blair RC and Higgins JJ (1981): A note on the asymptotic relative efficiency of the Wilcoxon rank-sum test relative to the independent means *t* test under mixtures of two normal distributions. *British Journal of Mathematical and Statistical Psychology* 34, 124–128.

Blair RC, Sawilowsky S (1993): Comparison of two tests useful in situations where treatment is expected to increase variability relative to controls. *Statistics in Medicine* 12, 2233–2243.

Blomqvist D, Andersson M, Küpper C, Cuthill IC, Kis J, Lanctot RB, Sandercock BK, Szekely T, Wallander J and Kempenaers B (2002): Genetic similarity between mates and extra-pair parentage in three species of shorebirds. *Nature* 419, 613–615.

Boik RJ (1987): The Fisher-Pitman permutation test: A non-robust alternative to the normal theory F test when variances are heterogeneous. *British Journal of Mathematical and Statistical Psychology* 40, 26–42.

Boneau CA (1960): The effects of violations of assumptions underlying the t test. *Psychological Bulletin* 57, 49–64.

Boos DD and Brownie C (1992): A rank-based mixed model approach to multisite clinical trials. *Biometrics* 48, 61–72.

Boschloo RD (1970): Raised conditional level of significance for the 2x2-table when testing the equality of two probabilities. *Statistica Neerlandica* 24, 1–35.

Box GEP (1953): Non-normality and tests on variances. *Biometrika* 40, 318–335.

Box GEP and Andersen SL (1955): Permutation theory in the derivation of robust criteria and the study of departures from assumption. *Journal of the Royal Statistical Society, Series B* 17, 1–26.

Bradley JV (1968): *Distribution-free statistical tests.* Prentice-Hall, Englewood Cliffs, NJ.

Bradley JV (1977) A common situation conducive to bizarre distribution shapes. *American Statistician* 31, 147–150.

Bretz F, Hothorn LA (2001): Testing dose-response relationships with a priori unknown, possibly non-monotone shapes. *Journal of Biopharmaceutical Statistics* 11, 193–207.

Brown MB and Forsythe AB (1974): Robust tests for the equality of variances. *Journal of the American Statistical Association* 69, 364–367.

Brownie C, Boos DD and Hughes-Oliver J (1990): Modifying the t and ANOVA F tests when treatment is expected to increase variability relative to controls. *Biometrics* 46, 259–266.

Brunner E, Dette H, Munk A (1997): Box-type approximations in nonparametric factorial designs. *Journal of the American Statistical Association* 92, 1494–1502.

Brunner E and Munzel U (2000): The nonparametric Behrens-Fisher problem: asymptotic theory and a small sample approximation. *Biometrical Journal* 42, 17–25.

Brunner E and Munzel U (2002): *Nichtparametrische Datenanalyse.* Springer, Berlin.

Buck W (1979): Signed-rank tests in the presence of ties (with extended tables). *Biometrical Journal* 21, 501–526.

Budde M and Bauer P (1989): Multiple test procedures in clinical dose finding studies. *Journal of the American Statistical Association* 84, 792–796.

Büning, H (1991): *Robuste und adaptive Tests.* De Gruyter, Berlin.

Büning, H (1996): Adaptive tests for the c-sample location problem–The case of two-sided alternatives. *Communications in Statistics—Theory and Methods* 25, 1569–1582.

Büning H (1997): Robust analysis of variance. *Journal of Applied Statistics* 24, 319–332.

Büning H (2002): Robustness and power of modified Lepage, Kolmogorov-Smirnov and Cramér-von Mises two-sample tests. *Journal of Applied Statistics* 29, 907–924.

Büning H and Trenkler G (1994): *Nichtparametrische statistische Methoden.* De Gruyter, Berlin (2nd edition).

Chen JJ, Kodell RL and Pearce BA (1997): Significance levels of randomization trend tests in the event of rare occurrences. *Biometrical Journal* 39, 327–337.

Chen RS and Dunlap WP (1993): SAS procedures for approximate randomization tests. *Behavior Research Methods, Instruments, and Computers* 25, 406–409.

Chen YI (1991): Notes on the Mack-Wolfe and Chen-Wolfe tests for umbrealla alternatives. *Biometrical Journal* 33, 281–290.

Chen YI, Wolfe DA (1990): A study of distribution-free tests for umbrella alternatives. *Biometrical Journal* 32, 47–57.

Coakley CW, Heise MA (1996): Versions of the sign test in the presence of ties. *Biometrics* 52, 1242–1251.

Cohen A and Sackrowitz HB (2003): Methods of reducing loss of efficiency due to discreteness of distributions. *Statistical Methods in Medical Research* 12, 23–36.

Corcoran C, Mehta C and Senchaudhuri P (2000): Power comparisons for tests of trend in dose-response studies. *Statistics in Medicine* 19, 3037–3050.

Cox DR (1958): Some problems connected with statistical inference. *Annals of Mathematical Statistics* 15, 357–372.

Cox DR (1977): The role of significance tests. *Scandinavian Journal of Statistics* 4, 49–70.

Crowley PH (1992): Resampling methods for computer-intensive data analysis in ecology and evolution. *Annual Review of Ecology and Systematics* 23, 405–427.

Cucconi O (1968): Un nuovo test non parametrico per il confronto tra due gruppi campionari. *Giornale degli Economisti* 27, 225–248 [cited according to Marozzi, 2008].

Cushman JH, Lawton JH and Manly BFJ (1993): Latitudinal patterns in European ant assemblages: Variation in species richness and body size. *Oecologia* 95, 30–37.

Cuzick J (1985): A Wilcoxon-type test for trend. *Statistics in Medicine* 4, 87–90.

Cytel Software Corporation (2007): *StatXact 8: User manual.* Cambridge.

Darwin C (1876): *The effect of cross- and self-fertilization in the vegetable kingdom.* John Murray, London (2nd edition).

Delaney HD and Vargha A (2002): Comparing several robust tests of stochastic equality with ordinally scaled variables and small to moderate sized samples. *Psychological Methods* 7, 485–503.

Demissie M, Mascialino B, Calza S and Pawitan Y (2008): Unequal group variances in microarray data analyses. *Bioinformatics* 24, 1168–1174.

Deuchler G (1914): Über die Methoden der Korrelationsrechnung in der Pädagogik und Psychologie. *Zeitschrift für pädagogische Psychologie und experimentelle Pädagogik* 15, 114–131 [cited according to Kruskal, 1957].

Duller C (2008): *Einführung in die nichtparametrische Statistik mit SAS und R.* Physica-Verlag, Heidelberg.

Duran BS (1976): A survey of nonparametric tests for scale. *Communications in Statistics—Theory and Methods* 5, 1287–1312.

Edgington ES and Onghena P (2007): *Randomization tests.* Chapman and Hall/CRC, Boca Raton, FL (4th edition).

Efron B (1982): *The jackknife, the bootstrap and other resampling plans.* Society for Industrial and Applied Mathematics, Philadelphia.

Efron B and Tibshirani RJ (1993): *An introduction to the bootstrap.* Chapman and Hall, New York.

EMEA (2003): *Points to consider on adjustment for baseline covariates.* European Agency for the Evaluation of Medicinal Products (EMEA), London.

Everitt BS and Hothorn T (2006): *A handbook of statistical analyses using R.* Chapman and Hall/CRC, Boca Raton, FL.

Fagerland MW and Sandvik L (2009a): Performance of five two-sample location tests for skewed distributions with unequal variances. *Contemporary Clinical Trials* 30, 490–496.

Fagerland MW and Sandvik L (2009b): The Wilcoxon-Mann-Whitney test under scrutiny. *Statistics in Medicine* 28, 1487–1497.

Fisher LD and van Belle G (1993): *Biostatistics.* Wiley, New York.

Fisher RA (1936): "The coefficient of racial likeness" and the future of craniometry. *Journal of the Royal Anthropological Institute* 66, 57–63 [reprint in Bennett JH (Ed.): Collected papers of R.A. Fisher, Vol. III. Adelaide, 1973, pp. 484–490].

Fligner MA and Policello GE (1981): Robust rank procedures for the Behrens-Fisher problem. *Journal of the American Statistical Association* 76, 162–168.

Fong DYT, Kwan CW, Lam KF and Lam KSL (2003): Use of the sign test for the median in the presence of ties. *American Statistician* 57, 237–240.

Francis RICC and Manly BFJ (2001): Bootstrap calibration to improve the reliability of tests to compare sample means and variances. *Environmetrics* 12, 713–729.

Freidlin B and Korn EL (2002): A testing procedure for survival data with few responders. *Statistics in Medicine* 21, 65–78.

Freidlin B, Miao W and Gastwirth JL (2003): On the use of the Shapiro-Wilk test in two-stage adaptive inference for paired data from moderate to very heavy tailed distributions. *Biometrical Journal* 45, 887–900.

Freidlin B, Podgor MV and Gastwirth JL (1999): Efficiency robust tests for survival or ordered categorical data. *Biometrics* 55, 883–886.

Freidlin B, Zheng G, Li Z and Gastwirth JL (2002): Trend tests for case-control studies of genetic markers: Power, sample size and robustness. *Human Heredity* 53, 146–152.

Games PA (1984): Data transformation, power, and skew: a rebuttal to Levine and Dunlap. *Psychological Bulletin* 95, 345–347.

Gastwirth JL (1966): On robust procedures. *Journal of the American Statistical Association* 61, 929–948.

Gastwirth JL (1970): On robust rank tests. In: Puri ML (Ed.) *Nonparametric techniques in statistical inference.* Cambridge University Press, Cambridge, pp. 89–109.

Gastwirth JL and Freidlin B (2000): On power and efficiency robust linkage tests for affected sibs. *Annals of Human Genetics* 64, 443–453.

Gebhard J (1995): *Optimalitätseigenschaften und Algorithmen für Permutationstests.* Skripten zur Mathematischen Statistik, Nr. 26 (reprint of PhD thesis), Münster.

Gefeller O and Bregenzer T (1994): Computer programs for exact nonparametric inference. *CABIOS—Computer Applications in the Biosciences* 10, 213–214.

George EO, Mudholkar DS (1990): P-values for two-sided tests. *Biometrical Journal* 32, 747–751.

Gibbons JD (1993): *Nonparametric statistics: an introduction.* Sage, Newbury Park, CA.

Göggelmann W (1993): Die Erfassung von Genmutationen in Bakterien. In: Fahrig R (Ed.): *Mutationsforschung und genetische Toxikologie.* Wissenschaftliche Buchgesellschaft, Darmstadt, pp. 207–216.

Good PI (2000): *Permutation tests.* Springer, New York (2nd edition).

Good PI (2001): *Resampling methods: a practical guide to resampling methods.* Birkhäuser, Boston (2nd edition).

Good PI (2005): *Introduction to statistics through resampling methods and R/S-Plus.* Wiley, Hoboken, NJ.

Gould SJ (1996): *Full house: the spread of excellence from Plato to Darwin.* Harmony Books, New York.

Gregoire TG and Driver BL (1987): Analysis of ordinal data to detect population differences. *Psychological Bulletin* 101, 159–165.

Hall P and Wilson SR (1991): Two guidelines for bootstrap hypothesis testing. *Biometrics* 47, 757–762.

Hall P and Yao Q (2003): Inference in ARCH and GARCH models with heavy-tailed errors. *Econometrica* 71, 285–317.

Hand DJ, Daly, F, Lunn, AD, McConway and Ostrowski E (1994): *A handbook of small data sets*. Chapman & Hall, London.

Hartung J, Elpelt B and Klösener KH (2009): *Statistik*. Oldenbourg, München (15th edition).

Hartung J, Knapp G and Sinha BK (2008): *Statistical meta-analysis with applications*. Wiley, Hoboken, NJ.

Hayes AF (2000): Randomization tests and the equality of variance assumption when comparing group means. *Animal Behaviour* 59, 653–656.

Hettmansperger TP and McKean JW (1998): *Robust nonparametric statistical methods*. Arnold, London.

Hettmansperger TP and Norton RM (1987): Tests for patterned alternatives in k-sample problems. *Journal of the American Statistical Association* 82, 292–299.

Higgins JJ (2004): *An introduction to modern nonparametric statistics*. Brooks/Cole, Pacific Grove, CA.

Hill NJ, Padmanabhan AR and Puri ML (1988): Adaptive nonparametric procedures and applications. *Applied Statistics* 37, 205–218.

Hines WGS and O'Hara Hines RJ (2000): Increased power with modified forms of the Levene (med) test for heterogeneity of variance. *Biometrics* 56, 451–454.

Hodges JL and Lehmann EL (1956): The efficiency of some nonparametric competitors of the *t*-test. *Annals of Mathematical Statistics* 27, 324–335.

Hoeffding W (1952): The large-sample power of tests based on permutations of observations. *Annals of Mathematical Statistics* 23, 169–192.

Hogg, RV (1974): Adaptive robust procedures: a partial review and some suggestions for future applications and theory. *Journal of the American Statistical Association* 69, 909–927.

Hogg RV, Fisher DM and Randles RH (1975): A two-sample adaptive distribution-free test. *Journal of the American Statistical Association* 70, 656–661.

Hollander M and Wolfe DA (1999): *Nonparametric statistical methods*. Wiley, New York (2nd edition).

Horn M (1990): Zum Test von Wilcoxon, Mann und Whitney: Bedingungen, unter denen und Fragestellungen, für die er anwendbar ist. *Zeitschrift für Versuchstierkunde* 33, 109–114.

Hothorn LA (1990): Biometrische Analyse spezieller Untersuchungen der regulativen Toxikologie. In: Klöcking HP, Güttner J and Wiezorek WD (eds.): *Grundlagen der Statistik für Toxikologen*. Verlag Gesundheit, Berlin, 2nd edition, pp. 130–238.

Hothorn LA and Hauschke D (1998): Principles in statistical testing in randomized toxicological studies. In: Chow SC and Liu JP (Eds.): *Designs and analysis of animal studies in pharmaceutical development*. Marcel Dekker, New York, pp. 79–133.

Hothorn T and Hornig K (2002): Exact nonparametric inference in R. In: Härdle W and Rönz B (Eds.): *Compstat: Proceedings in Computational Statistics, 15th Symposium*. Physica-Verlag, Heidelberg, pp. 355–360.

Hothorn T, Hornik K, van de Wiel MA and Zeileis A (2006): A lego system for conditional inference. *American Statistician* 60, 257–263.

Huang Y, Xu H, Calian V and Hsu JC (2006): To permute or not to permute. *Bioinformatics* 22, 2244–2248.

Hunter MA and May RB (1993): Some myths concerning parametric and nonparametric tests. *Canadian Psychology* 34, 384–389.

ICH (1999): ICH harmonized tripartite guideline E9: Statistical principles for clinical trials. *Statistics in Medicine* 18, 1905–1942.

Jansen RC (2001): Quantitative trait loci in inbred lines. In: Balding DJ, Bishop M and Cannings C (Eds.): *Handbook of statistical genetics*. Wiley, Chichester, pp. 567–597.

Janssen A (1997): Studentized permutation tests for non-i.i.d. hypotheses and the generalized Behrens-Fisher problem. *Statistics and Probability Letters* 36, 9–21.

Janssen A (1998): *Zur Asymptotik nichtparametrischer Tests*. Skripten zur Mathematischen Statistik, Nr. 29, Münster.

Janssen A and Pauls T (2003): How do bootstrap and permutation tests work? *Annals of Statistics* 31, 768–806.

Jonckheere AR (1954): A distribution-free k-sample test against ordered alternatives. *Biometrika* 41, 133–145.

Kasuya E (2001): Mann-Whitney U test when variances are unequal. *Animal Behaviour* 61, 1247–1249.

Keller-McNulty S and Higgins JJ (1987): Effect of tail weight on power and type-I error of robust permutation tests for location. *Communications in Statistics—Simulation and Computation* 16, 17–35.

Kennedy PE (1995): Randomization tests in econometrics. *Journal of Business and Economic Statistics* 13, 85–94.

Keyes TK and Levy MS (1997): Analysis of Levene's test under design imbalance. *Journal of Educational and Behavioral Statistics* 22, 227–236.

Klar B, Petney TN and Taraschewski H (2010): Quantifying differences in parasite numbers between samples of hosts. *Journal of Parasitology* 96, 856–861.

Knijnenburg TA, Wessels LFA, Reinders, MJT and Shmulevich I (2009): Fewer permutations, more accurate P-values. *Bioinformatics* 25, i161–i168.

Krackow S and Tkadlec E (2001): Analysis of brood sex ratios: implications of offspring clustering. *Behavioral Ecology and Sociobiology* 50, 293–301.

Krämer W (1992): *Statistik verstehen: eine Gebrauchsanweisung*. Campus, Frankfurt/Main.

Kropf S (1988): Application of multiple test procedures to the combination of multivariate and univariate tests with varying variable sets. *Biometrical Journal* 30, 461–470.

Kruskal WH (1957): Historical notes on the Wilcoxon unpaired two-sample test. *Journal of the American Statistical Association* 52, 356–360.

Labovitz S (1970): The assignment of numbers to rank order categories. *American Sociological Review* 35, 515–524.

Lachenbruch PA (1976): Analysis of data with clumping at zero. *Biometrische Zeitschrift* 18, 351–356.

Lachenbruch PA (2002): Analysis of data with excess zeros. *Statistical Methods in Medical Research* 11, 297–302.

Lancaster HO (1961): Significance tests in discrete distributions. *Journal of the American Statistical Association* 56, 223–234.

Larocque D and Randles RH (2008): Confidence intervals for a discrete population median. *American Statistician* 62, 32–39.

Le CT (1988): A new rank test against ordered alternatives in k-sample problems. *Biometrical Journal* 30, 87–92.

Le CT (1994): Some tests for linear trend of variances. *Communications in Statistics—Theory and Methods* 23, 2269–2282.

Leber PD and Davis CS (1998): Threats to the validity of clinical trials employing enrichment strategies for sample selection. *Controlled Clinical Trials* 19, 178–187.

Lehmacher W (1976): *Asymptotische Eigenschaften linearer Zweistichproben-Rangtests bei beliebigen Verteilungen.* PhD thesis, University of Dortmund.

Lehmacher W and Wassmer G (1999): Adaptive sample size calculations in group sequential trials. *Biometrics* 55, 1286–1290.

Lehmann EL (2006): *Nonparametrics: Statistical methods based on ranks.* Springer, New York (revised first edition).

Lehmann EL (2009): Parametric versus nonparametrics: Two alternative methodologies. *Journal of Nonparametric Statistics* 21, 397–405.

Lehmann EL and Stein C (1949): On the theory of some non-parametric hypotheses. *Annals of Mathematical Statistics* 20, 28–45.

Lepage Y (1971): A combination of Wilcoxon's and Ansari-Bradley's statistics. *Biometrika* 58, 213–217

Leuchs AK and Neuhäuser M (2010): A SAS/IML algorithm for exact nonparametric paired tests. *GMS Medizinische Informatik, Biometrie und Epidemiologie* 6, Doc4.

Levene H (1960): Robust tests for equality of variances. In: Olkin I, Ghurye SG, Hoeffding W, Madow WG and Mann HB (eds.): *Contributions to probability and statistics.* Stanford University Press, Stanford, pp. 278–292.

Lin KK and Ali MW (1994): Statistical review and evaluation of animal tumorigenicity studies. In: Buncher CR and Tsay JY (eds.): *Statistics in the pharmaceutical industry.* Dekker, New York, pp. 19–57.

Liu X, Nickel R, Beyer K, Wahn U, Ehrlich E, Freidhoff LR, Björksten B, Beaty TH, Huang SK, and the MAS-Study Group (2000): An IL13 coding region variant is associated with a high total serum IgE level and atopic dermatitis in the German multicenter atopy study (MAS-90). *Journal of Allergy and Clinical Immunology* 106, 167–170.

Lock RH (1991): A sequential approximation to a permutation test. *Communications in Statistics—Simulation and Computation* 20, 341–363.

Lu Y (2010): A revised version of McNemar's test for paired binary data. *Communications in Statistics—Theory and Methods* 39, 3525–3539.

Ludbrook J and Dudley H (1994): Issues in biomedical statistics: Statistical inference. *Australian and New Zealand Journal of Surgery* 64, 630–636.

Ludbrook J and Dudley H (1998): Why permutation tests are superior to t and F tests in biomedical research. *American Statistician* 52, 127–132.

Lüdin E (1985): A test procedure based on ranks for the statistical evaluation of toxicological studies. *Archives of Toxicology* 58, 57–58.

Lydersen S, Fagerland MW and Laake P (2009): Recommended tests for association in 2x2 tables. *Statistics in Medicine* 28, 1159–1175.

Magel RC (1986): A comparison of some nonparametric tests for small sample sizes. *Proceedings of the Modeling and Simulation Conference 17, Part V.* (zitiert nach Mahrer and Magel, 1995)

Mahrer JM and Magel RC (1995): A comparison of tests for the k-sample, non-decreasing alternative. *Statistics in Medicine* 14, 863–871.

Malik HJ (1985): Logistic distribution. In: Kotz, S. and Johnson, N. L. (eds.): *Encyclopedia of statistical sciences, Vol. 5.* Wiley, New York, pp. 123–128.

Manly BFJ (1995): Randomization tests to compare means with unequal variation. *Sankhya B* 57, 200–222.

Manly BFJ (2007): *Randomization, bootstrap and Monte Carlo methods in biology.* Chapman & Hall/CRC, London (3rd edition).

Manly BFJ and Francis RICC (1999): Analysis of variance by randomization when variances are unequal. *Australian and New Zealand Journal of Statistics* 41, 411–429.

Manly BFJ and Francis RICC (2002): Testing for mean and variance differences with samples from distributions that may be non-normal with unequal variances. *Journal of Statistical Computation and Simulation* 72, 633–646.

Mann HB and Whitney DR (1947): On a test of whether one of two random variables is stochastically larger than the other. *Annals of Mathematical Statistics* 18, 50–60.

Marozzi M (2008): The Lepage location-scale test revisited. *Far East Journal of Theoretical Statistics* 24, 137–155.

May RB and Hunter MA (1993): Some advantages of permutation tests. *Canadian Psychology* 34, 401–407.

Mayhew PJ and Pen I (2002): Comparative analysis of sex ratios. In: Hardy ICW (ed.): *Sex ratios: concepts and research methods*. Cambridge University Press, Cambridge, pp. 132–156.

McArdle BH and Anderson MJ (2004): Variance heterogeneity, transformations, and models of species abundance: a cautionary tale. *Canadian Journal of Fisheries and Aquatic Science* 61, 1294–1302.

Mehrotra DV, Chan ISF and Berger RL (2003): A cautionary note on exact unconditional inference for a difference between two independent binomial proportions. *Biometrics* 59, 441–450.

Mehta CR and Hilton JF (1993): Exact power of conditional and unconditional tests: Going beyond the 2x2 contingency table. *American Statistician* 47, 91–98.

Mehta CR and Patel N (1983): A network algorithm for performing Fisher's exact test in rxc contingency tables. *Journal of the American Statistical Association* 78, 427–434.

Mehta CR, Patel N and Senchaudhuri P (1992): Exact stratified linear rank tests for ordered categorical and binary data. *Journal of Computational and Graphical Statistics* 1, 21–40.

Micceri T (1989): The unicorn, the normal curve, and other improbable creatures. *Psychological Bulletin* 105, 156–166.

Mielke HW, Gonzales CR, Smith MK and Mielke PW (1999): The urban environment and children's health: Soils as an indicator of lead, zinc, and cadmium in New Orleans, Louisiana, U.S.A. *Environmental Research (Section A)* 81, 117–129.

Mielke PW, Johnston JE and Berry KJ (2004): Combining probability values from independent permutation tests: a discrete analog of Fisher's classical method. *Psychological Reports* 95, 449–458.

Mood AM, Graybill, FA and Boes DC (1974): *Introduction to the theory of statistics*. McGraw-Hill, New York (3rd edition).

Mundry R and Fischer J (1998): Use of statistical programs for nonparametric tests of small samples often leads to incorrect P values: Examples from Animal Behaviour. *Animal Behaviour* 56, 256–259.

Munzel U and Brunner E (2002): An exact paired rank test. *Biometrical Journal* 44, 584–593.

Murakami H (2006): A *K*-sample rank test based on a modified Baumgartner statistic and its power comparison. *Journal of the Japanes Society of Computational Statistics* 19, 1–13.

Murakami H (2007): Lepage type statistic based on the modified Baumgartner statistic. *Computational Statistics and Data Analysis* 51, 5061–5067.

Murakami H (2008): A multisample rank test for location-scale parameters. *Communications in Statistics—Simulation and Computation* 37, 1347–1355.

Nanna MJ and Sawilowsky SS (1998): Analysis of Likert scale data in disability and medical rehabilitation research. *Psychological Methods* 3, 55–67.

Neubert K (2006): *Das nichtparametrische Behrens-Fisher-Problem: ein studentisierter Permutationstest und robuste Konfidenzintervalle für den Shift-Effekt.* PhD thesis, University of Göttingen.

Neubert K and Brunner E (2007): A studentized permutation test for the nonparametric Behrens-Fisher problem. *Computational Statistics and Data Analysis* 51, 5192–5204.

Neuhäuser M (2000): An exact two-sample test based on the Baumgartner-Weiss-Schindler statistic and a modification of Lepage's test. *Communications in Statistics—Theory and Methods* 29, 67–78.

Neuhäuser M (2001a): One-sided two-sample and trend tests based on a modified Baumgartner-Weiss-Schindler statistic. *Journal of Nonparametric Statistics* 13, 729–739.

Neuhäuser M (2001b): An adaptive location-scale test. *Biometrical Journal* 43, 809–819.

Neuhäuser M (2001c): An adaptive interim analysis – a useful tool for ecological studies. *Basic and Applied Ecology* 2, 203–207.

Neuhäuser M (2002a): Nonparametric identification of the minimum effective dose. *Drug Information Journal* 36, 881–888.

Neuhäuser M (2002b): The Baumgartner-Weiss-Schindler test in the presence of ties (letter to the editor). *Biometrics* 58, 250.

Neuhäuser M (2002c): Two-sample tests when variances are unequal. *Animal Behaviour* 63, 823–825.

Neuhäuser M (2002d): Exact tests for the analysis of case-control studies of genetic markers. *Human Heredity* 54, 151–156.

Neuhäuser M (2003a): Tests for genetic differentiation. *Biometrical Journal* 45, 974–984.

Neuhäuser M (2003b): *Nichtparametrische Zweistichprobentests bei potentiell ungleichen Varianzen.* Habilitation thesis, University of Dortmund.

Neuhäuser M (2003c): A note on the exact test based on the Baumgartner-Weiss-Schindler statistic in the presence of ties. *Computational Statistics and Data Analysis* 42, 561–568.

Neuhäuser M (2003d): Further evidence for Emlen's hypothesis from two parrot species. *New Zealand Journal of Zoology* 30, 221–225.

Neuhäuser M (2004): Wilcoxon test after Levene's transformation can have an inflated type I error rate. *Psychological Reports* 94, 1419–1420.

Neuhäuser M (2005a): Exact tests based on the Baumgartner-Weiss-Schindler statistic—A survey. *Statistical Papers* 46, 1–30.

Neuhäuser M (2005b): One-sided nonparametric tests for ordinal data. *Perceptual and Motor Skills* 101, 510–514.

Neuhäuser M (2006a): An exact test for trend among binomial proportions based on a modified Baumgartner-Weiss-Schindler statistic. *Journal of Applied Statistics* 33, 79–88.

Neuhäuser M (2006b): How to deal with multiple endpoints in clinical trials. *Fundamental and Clinical Pharmacology* 20, 515–523.

Neuhäuser M (2007): A comparative study of nonparametric two-sample tests after Levene's transformation. *Journal of Statistical Computation and Simulation* 77, 517–526.

Neuhäuser M (2010): A nonparametric two-sample comparison for skewed data with unequal variances. *Journal of Clinical Epidemiology* 63, 691–693.

Neuhäuser M (2011): Transformations can be avoided when comparing skewed distributions with unequal variances (letter to the editor). *Journal of Clinical Epidemiology* 64, 454–455.

Neuhäuser M, Boes T and Jöckel KH (2005): Two-part permutation tests for DNA methylation and microarray data. *BMC Bioinformatics* 6, 35.

Neuhäuser M, Boes T and Jöckel KH (2007): Pseudo-precision in gene expression values can reduce efficiency. *Methods of Information in Medicine* 46, 538–541.

Neuhäuser M and Bretz F (2001): Nonparametric all-pairs multiple comparisons. *Biometrical Journal* 43, 571–580.

Neuhäuser M, Büning H and Hothorn LA (2004): Maximum test versus adaptive tests for the two-sample location problem. *Journal of Applied Statistics* 31, 215–227.

Neuhäuser M and Hothorn LA (1998): An analogue of Jonckheere's trend test for parametric and dichotomous data. *Biometrical Journal* 40, 11–19.

Neuhäuser M and Hothorn LA (1999): An exact Cochran-Armitage test for trend when dose-response shapes are a priori unknown. *Computational Statistics and Data Analysis* 30, 403–412.

Neuhäuser M and Hothorn LA (2000): Parametric location-scale and scale trend tests based on Levene's transformation. *Computational Statistics and Data Analysis* 33, 189–200.

Neuhäuser M and Hothorn LA (2006): Maximum tests are adaptive permutation tests. *Journal of Modern Applied Statistical Methods* 5, 317–322.

Neuhäuser M and Jöckel K-H (2006): A bootstrap test for the analysis of microarray experiments with a very small number of replications. *Applied Bioinformatics* 5, 173–179.

Neuhäuser M and Lam FC (2004): Nonparametric approaches to detecting differentially expressed genes in replicated microarray experiments. In: Chen, Y.-P.P. (Ed.): *Conferences in Research and Practice in Information Technology*, Vol. 29, pp. 139–143 (Proceedings of the 2nd Asia-Pacific Bioinformatics Conference). Australian Computer Society, Adelaide.

Neuhäuser M, Leisler B and Hothorn LA (2003): A trend test for the analysis of multiple paternity. *Journal of Agricultural, Biological and Environmental Statistics* 8, 29–35.

Neuhäuser M, Liu PY and Hothorn LA (1998): Nonparametric tests for trend: Jonckheere's test, a modification and a maximum test. *Biometrical Journal* 40, 899–909.

Neuhäuser M and Manly BFJ (2004): The Fisher-Pitman permutation test when testing for differences in mean and variance. *Psychological Reports* 94, 189–194.

Neuhäuser M and Poulin R (2004): Comparing parasite numbers between samples of hosts. *Journal of Parasitology* 90, 689–691.

Neuhäuser M and Ruxton GD (2009a): Round your numbers in rank tests: exact and asymptotic inference and ties. *Behavioral Ecology and Sociobiology* 64, 297–303.

Neuhäuser M and Ruxton GD (2009b): Distribution-free two-sample comparisons in the case of heterogeneous variances. *Behavioral Ecology and Sociobiology* 63, 617–623.

Neuhäuser M, Schulz A and Czech D (2009): A SAS/IML algorithm for an exact permutation test. *GMS Medizinische Informatik, Biometrie und Epidemiologie* 5, Doc13.

Neuhäuser M, Seidel D, Hothorn LA and Urfer W (2000): Robust trend tests with application to toxicology. *Environmental and Ecological Statistics* 7, 43–56.

Neuhäuser M and Senske R (2004): The Baumgartner-Weiss-Schindler test for the detection of differentially expressed genes in replicated microarray experiments. *Bioinformatics* 20, 3553–3564.

Neuhäuser M and Senske R (2009): The analysis of multicentre clinical trials when there is heterogeneity between centres. *Journal of Statistical Computation and Simulation* 79, 1381–1387.

Noether GE (1987): Sample size determination for some common nonparametric tests. *Journal of the American Statistical Association* 82, 645–647.

North BV, Curtis D, Sham PC (2002): A note on the calculation of empirical P values from Monte Carlo procedures. *American Journal of Human Genetics* 71, 439–441.

O'Brien PC (1988): Comparing two samples: Extensions to the *t*, rank-sum and log-rank tests. *Journal of the American Statistical Association* 83, 52–61.

Ogenstad, S (1998): The use of generalized tests in medical research. *Journal of Biopharmaceutical Statistics* 8, 497–508.

O'Neill ME and Mathews K (2000): A weighted least squares approach to Levene's test of homogeneity of variance. *Australian and New Zealand Journal of Statistics* 42, 81–100.

Onghena P and May RB (1995): Pitfalls in computing and interpreting randomization test p values: A commentary on Chen and Dunlap. *Behavior Research Methods, Instruments, and Computers* 27, 408–411.

Opdyke JD (2003): Fast permutation tests that maximize power under conventional Monte Carlo sampling for pairwise and multiple comparisons. *Journal of Modern Applied Statistical Methods* 2, 27–49.

Pan G (2002): Confidence intervals for comparing two scale parameters based on Levene statistics. *Journal of Nonparametric Statistics* 14, 459–476.

Pepe MS, Longton G, Anderson GL and Schummer M (2003): Selecting differentially expressed genes from microarray experiments. *Biometrics* 59, 133–142.

Pesarin F (2001): *Multivariate permutation tests*. Wiley, New York.

Pesarin F and Salmaso L (2010): *Permutation tests for complex data: Theory, applications and software*. Wiley, New York.

Pettitt AN (1976): A two-sample Anderson-Darling rank statistic. *Biometrika* 63, 161–168.

Piegorsch WW and Bailer AJ (1997): *Statistics for environmental biology and toxicology*. Chapman & Hall, London.

Pigeot I (2000): Basic concepts of multiple tests—A survey. *Statistical Papers* 41, 3–36.

Pitman EJG (1937): Significance tests which may be applied to samples from any populations. *Supplement to the Journal of the Royal Statistical Society* 4, 119–130.

Pratt JW (1959): Remarks on zeros and ties in the Wilcoxon signed rank procedures. *Journal of the American Statistical Association* 54, 655–667.

Proschan MA and Nason M (2009): Conditioning in 2x2 tables. *Biometrics* 65, 316–322.

Putter J (1955): The treatment of ties in some nonparametric tests. *Annals of Mathematical Statistics* 26, 368–386.

Rabbee N, Coull BA, Mehta C, Patel N and Senchaudhuri P (2003): Power and sample size for ordered categorical data. *Statistical Methods in Medical Research* 12, 73–84.

Rahlfs VW and Zimmermann H (1993): Scores: ordinal data with few categories—How they should be analyzed. *Drug Information Journal* 27, 1227–1240.

Randles RH and Wolfe DA (1979): *Introduction to the theory of nonparametric statistics*. Wiley, New York.

Rasch D and Verdooren R (2004): *Grundlagen der Korrelationsanalyse und der Regressionsanalyse.* Saphir Verlag, Ribbesbüttel.

Rasmussen JL (1986): An evaluation of parametric and non-parametric tests on modified and non-modified data. *British Journal of Mathematical and Statistical Psychology* 39, 213–220.

Reiczigel J, Zakarias I and Rózsa L (2005): A bootstrap test of stochastic equality of two populations. *American Statistician* 59, 156–161.

Reiser B and Guttman I (1986): Statistical inference for $\Pr(Y < X)$: The normal case. *Technometrics* 28, 253–257.

Rice WR (1990): A consensus combined p-value test and the family-wide significance of component tests. *Biometrics* 46, 303–308.

Rice WR and Gaines SD (1989): One-way analysis of variance with unequal variances. *Proceedings of the National Academy of Sciences USA* 86, 8183–8184.

Rodgers JL (1999): The bootstrap, the jackknife, and the randomization test: A sampling taxonomy. *Multivariate Behavioral Research* 34, 441–456.

Romano JP (1989): Bootstrap and randomization tests of some nonparametric hypotheses. *Annals of Statistics* 17, 141–159.

Romano JP (1990): On the behavior of randomization tests without a group invariance assumption. *Journal of the American Statistical Association* 85, 686–692.

Rorden C, Bonilha L, Nichols TE (2007): Rank-order versus mean based statistics for neuroimaging. *NeuroImage* 35, 1531–1537.

Rosner B and Glynn RJ (2009): Power and sample size estimation for the Wilcoxon rank sum test with application to comparisons of C statistics from alternative prediction models. *Biometrics* 65, 188–197.

Rózsa L, Reiczigel J and Majoros G (2000): Quantifying parasites in samples of hosts. *Journal of Parasitology* 86, 228–232.

Ruberg SJ (1995): Dose response studies. II. Analysis and interpretation. *Journal of Biopharmaceutical Statistics* 5, 1542.

Rüther E, Degner D, Munzel U, Brunner E, Lenhard G, Biehl J and Vögtle-Junkert U (1999): Antidepressant action of sulpiride. Results of a placebo-controlled double-blind trial. *Pharmacopsychiatry* 32, 127–135.

Ruxton GD and Neuhäuser M (2010a): When should we use one-tailed hypothesis testing? *Methods in Ecology and Evolution* 1, 114–117.

Ruxton GD and Neuhäuser M (2010b): Good practice in testing for an association in contingency tables. *Behavioral Ecology and Sociobiology* 64, 1505–1513.

Ruxton GD, Rey D and Neuhäuser M (2010): Comparing samples with large numbers of zeros. *Animal Behaviour* 80, 937–940.

Ryman N and Jorde PE (2001): Statistical power when testing for genetic differentiation. *Molecular Ecology* 10, 2361–2373.

Saino N, Ellegren H and Moller AP (1999): No evidence for adjustment of sex allocation in relation to paternal ornamentation and paternity in barn swallows. *Molecular Ecology* 8, 399–406.

SAS Institute Inc. (2004): SAS/STAT 9.1 User's Guide. SAS Institute Inc., Cary.

Sasieni PD (1997): From genotypes to genes: doubling the sample size. *Biometrics* 53, 1253–1261.

Sawilowsky SS and Blair RC (1992): A more realistic look at the robustness and type II error properties of the *t* test to departures from population normality. *Psychological Bulletin* 111, 352–360.

Schröer G and Trenkler D (1995): Exact and randomization distributions of Kolmogorov-Smirnov tests two or three samples. *Computational Statistics and Data Analysis* 20, 185–202.

Schultz B (1983): On Levene's test and other statistics of variation. *Evolutionary Theory* 6, 197–203.

Schulze-Hagen K, Swatschek I, Dyrcz A and Wink M (1993): Multiple Vaterschaften in Bruten des Seggenrohrsängers *Acrocephalus paludicola*: Erste Ergebnisse des DNA-Fingerprintings. *Journal für Ornithologie* 134, 145–154.

Sedlmeier P, Renkewitz F (2008): *Forschungsmethoden und Statistik in der Psychologie*. Pearson Studium, München.

Senn S (2007): Drawbacks to noninteger scoring for ordered categorical data. *Biometrics* 63, 296–298.

Sham P (1998): *Statistics in human genetics*. Arnold, London.

Sherratt TN and Wilkinson DM (2009): *Big questions in ecology and evolution*. Oxford University Press, Oxford.

Sheu CF (2002): Fitting mixed-effects models for repeated ordinal outcomes with the NLMIXED procedure. *Behavior Research Methods, Instruments, and Computers* 34, 151–157.

Shieh G, Jan SL and Randles RH (2006): On power and sample size determination for the Wilcoxon-Mann-Whitney test. *Journal of Nonparametric Statistics* 18, 33–43.

Shoemaker LH (1995): Tests for differences in dispersion based on quantiles. *American Statistician* 49, 179–182.

Shoemaker LH (2003): Fixing the *F* test for equal variances. *American Statistician* 57, 105–114.

Shoetake T (1981): Population genetical study of natural hybridization between *Papio anubis* and *Papio hamadryas*. *Primates* 22, 285–308.

Siegel S (1956): *Nonparametric statistics for the behavioral sciences*. McGraw-Hill, New York.

Siegmund KD, Laird PW and Laird-Offringa IA (2004): A comparison of cluster analysis methods using DNA methylation data. *Bioinformatics* 20, 1896–1904.

Singer, J (2001): A simple procedure to compute the sample size needed to compare two independent groups when the population variances are unequal. *Statistics in Medicine* 20, 1089–1095.

Slager SL and Schaid DJ (2001): Case-control studies of genetic markers: Power and sample size approximations for Armitage's test for trend. *Human Heredity* 52, 149–153.

Sokal RR and Braumann CA (1980): Significance tests for coefficients of variation and variability profiles. *Systematic Zoology* 29, 50–66.

Sokal RR and Rohlf FJ (1995): *Biometry*. W.H. Freeman and Company, New York (3rd edition).

Sprent P and Smeeton NC (2001): *Applied nonparametric statistical methods*. Chapman & Hall/CRC, Boca Raton, FL.

Steger H and Püschel F (1960): Der Einfluss der Feuchtigkeit auf die Haltbarkeit des Carotins in künstlich getrocknetem Grünfutter. *Die Deutsche Landwirtschaft* 11, 301–303.

Streitberg B and Röhmel J (1987): Exakte Verteilungen für Rang- und Randomisierungstests im allgemeinen c-Stichprobenproblem. *EDV in Medizin und Biologie* 18, 12–19.

Streitberg B and Röhmel J (1990): On tests that are uniformly more powerful than the Wilcoxon-Mann-Whitney test. *Biometrics* 46, 481–484.

Talwar PP and Gentle JE (1977): A robust test for the homogeneity of scales. *Communications in Statistics—Theory and Methods* 6, 363–369.

Tanizaki H. (1997): Power comparison of non-parametric tests: small-Sample properties from Monte Carlo experiments. *Journal of Applied Statistics* 24, 603–632.

Terpstra TJ (1952): The asymptotic normality and consistency of Kendall's test against trend, when ties are present in one ranking. *Indagationes Mathematicae* 14, 327–333.

ter Braak CJF (1992): Permutation versus bootstrap significance tests in multiple regression and ANOVA. In: Jöckel K-H, Rothe G and Sendler W (Eds.): *Bootstrapping and related techniques.* Springer, Berlin, pp. 79–85.

Thangavelu K and Brunner E (2007): Wilcoxon-Mann-Whitney test for stratified samples and Efron's paradox dice. *Journal of Statistical Planning and Inference* 137, 720–737.

Thomas F and Poulin R (1997): Using randomization techniques to analyse fluctuating asymmetry data. *Animal Behaviour* 54, 1027–1029.

Tilquin P, van Keilegom I, Coppieters W, le Boulenge E and Baret PV (2003): Non-parametric interval mapping in half-sib designs: Use of midranks to account for ties. *Genetical Research* 81, 221–228.

Tryon PV and Hettmansperger TP (1973): A class of non-parametric tests for homogeneity against ordered alternatives. *Annals of Statistics* 1, 1061–1070.

Tukey JW (1993): Tightening the clinical trial. *Controlled Clinical Trials* 14, 266–285.

van den Brink WP and van den Brink SGJ (1989): A comparison of the power of the *t* test, Wilcoxon's test, and the approximate permutation test for the two-sample location problem. *British Journal of Mathematical and Statistical Psychology* 42, 183–189.

van de Wiel MA and Di Bucchianico A (2001): Fast computation of the exact null distribution of Spearman's ρ and Page's L statistic for samples with and without ties. *Journal of Statistical Planning and Inference* 92, 133–145.

van Elteren PH (1960): On the combination of independent two sample tests of Wilcoxon. *Bulletin de l'Institut International de Statistique* 37, 351–361.

van Valen L (2005): The statistics of variation. In: Hallgrimsson B and Hall BK (Eds.): *Variation*. Elsevier, Amsterdam, pp. 29–47.

Vickers AJ (2005): Parametric versus non-parametric statistics in the analysis of randomized trials with on-normally distributed data. *BMC Medical Research Methodology* 5, 35.

Wang M, Matern B, Dmoch R, Neurohr B, Linke K and Schreiber A (1997): Erhaltungszuchten als Modelle genetischer Artenschutzprobleme: Das Beispiel dreier Primatenkolonien. In: Schreiber A and Lehmann J (Eds.): *Populationsgenetik im Artenschutz*. Landwirtschaftsverlag, Münster, pp. 153–169.

Weerahandi S (1995): *Exact statistical methods for data analysis*. Springer, New York.

Welch BL (1937): The significance of the difference between two means when the population variances are unequal. *Biometrika* 29, 350–362.

Weller EA and Ryan LM (1998): Testing for trend with count data. *Biometrics* 54, 762–773.

Westfall PH and Soper KA (1994): Nonstandard uses of proc multtest: Permutational Peto tests, permutational and unconditional *t* and binomial tests. *Proceedings of the 19th Annual SAS Users Group International Conference*. SAS Institute Inc., Cary, pp. 986–989.

Westfall PH and Young SS (1993): *Resampling-based multiple testing*. Wiley, New York.

Whitlock MC and Schluter D (2009): *The analysis of biological data*. Roberts, Greenwood Village, Co.

Wilcox RR (2003): *Applying contemporary statistical techniques*. Elsevier Academic Press, San Diego.

Wilcoxon F (1945): Individual comparisons by ranking methods. *Biometrics* 1, 80–83.

Williams DA (1988): Tests for differences between several small proportions. *Applied Statistics* 37, 421–434.

Williams PB and Carnine DW (1981): Relationship between range of examples and of instructions and attention in concept attainment. *Journal of Educational Research* 74, 144–148.

Wilson JB (2007): Priorities in statistics, the sensitive feet of elephants and dont transform data. *Folia Geobotanica* 42, 161–167.

Yang JJ (2010): Distribution of Fisher's combination statistic when the tests are dependent. *Journal of Statistical Computation and Simulation* 80, 1–12.

Yezerinac SM, Weatherhead PJ and Boag PT (1995): Extra-pair paternity and the opportunity for sexual selection in a socially monogamous bird (*Dendroica petechia*). *Behavioral Ecology and Sociobiology* 37, 179–188.

Zar JH (2010): *Biostatistical analysis*. Pearson Prentics Hall, Upper Saddle River, NJ (5th edition).

Zhang J (2006): Powerful two-sample tests based on the likelihood ratio. *Technometrics* 48, 95–103.

Zheng G, Freidlin B and Gastwirth JL (2002): Robust TDT-type candidate-gene association tests. *Annals of Human Genetics* 66, 145–155.

Zhou XH (2005): Nonparametric confidence intervals for the one- and two-sample problems. *Biostatistics* 6, 187–200.

Zimmerman DW (2003): A warning about the large-sample Wilcoxon-Mann-Whitney test. *Understanding Statistics* 2, 267–280.

Zöfel P (1992): *Univariate Varianzanalysen*. G. Fischer, Stuttgart.

Index

adaptive test, 36–38, 40, 42
Ansari-Bradley test, 61
anticonservative test, 191
approximate permutation test, 12, 13, 28, 77
Armitage test, 100, 101, 123, 124, 173
asymptotic relative efficiency, 7, 15, 32

back-up statistic, 111
Behrens-Fisher problem, 58, 68, 77, 84, 92
binomial test, 128
Bonferroni adjustment, 42, 192
Bonferroni-Holm procedure, 192
bootstrap, 73, 186
bootstrap t test, 74, 78
bootstrap calibration, 82
bootstrap confidence interval, 186
bootstrap estimator, 186
bootstrap methods, 109
bootstrap test, 73, 76, 77, 80, 84, 134, 136, 175, 176, 198
Brunner-Munzel test, 70–72, 85, 92, 198
BWS test, 22, 24, 25, 35, 97, 108, 195

case-control study, 121
closed testing procedure, 63, 66, 193
Cochran-Armitage test, 100, 123
combination test, 167, 169
completely randomized design, 139
complex designs, 175, 176
conditional test, 9, 100, 112

conservatism, 109
conservative test, 191
contingency table, 101, 155
continuity correction, 20
correlation coefficient, 158, 161
Cucconi test, 66–68, 92

D.O test, 81, 84, 116
discrete numerical data, 99, 101
distribution-free, 3

empirical distribution function, 93
exact χ^2 test, 104
exact permutation test, 12, 14
exchangeability, 44

false discovery rate, 193
finite relative efficiency, 7
Fisher's combination test, 64, 168–171
Fisher-Pitman permutation test, 7, 11, 14, 44, 142, 194
FPP test, 14, 74, 100, 101
Friedman test, 150–153

general alternative, 93

heteroscedasticity, 55, 58
Hodges-Lehmann estimator, 181, 183
homoscedasticity, 2

independence, 155
interim analysis, 64
interval scale, 189
inverse normal method, 168–170

jackknife, 186, 187